Practice Exam for the Civil PE Examination

Breadth + Construction Depth

Indranil Goswami, Ph.D., P.E.

September 2015

Second printing September 2015

Preface

In January 2015, the official (NCEES) syllabus for the PE-Civil examination underwent a significant realignment. There was a significant departure from the previous structure which placed approximately equal emphasis on the five areas of practice. In the breadth exam, the 40 questions were approximately equally distributed among Construction, Geotechnical, Structural, Transportation and Water Resources. In the current syllabus for the Breadth (AM) exam, Transportation has been significantly deemphasized while there seems to be more emphasis on Construction.

The new depth (PM) syllabi have also gone through reorganization as well as addition of specific subtopics under various categories.

These practice exams were developed *after* the syllabus went through the aforementioned reorganization and are therefore consistent with the same.

This full-length practice exam contains 40 breadth (AM) questions + 40 depth (PM) questions in the area of CONSTRUCTION ENGINEERING. It should be taken under as near exam conditions as possible, preferably at the point when you think your exam review is complete and you are ready to take a simulated test to assess the level of your preparation. You should even go so far as to ask someone else to detach the questions from the solutions, so that you don't have any temptation to peek.

All the best for the upcoming PE exam,

Indranil Goswami

P.S. In this second printing, errors discovered to date have been corrected.

Table of Contents

BREADTH EXAM

FOR THE

CIVIL PE EXAM

The following set of 40 questions (numbered 001 to 040) is representative of a 4-hour breadth (AM) exam according to the syllabus and guidelines for the Principles & Practice (P&P) of Civil Engineering Examination (updated January 2015) administered by the National Council of Examiners for Engineering and Surveying (NCEES). The exam is weighted according to the official NCEES syllabus (2015) in the following subject areas – Construction, Geotechnical, Structural, Transportation and Water & Environmental.

The time allocated for this set of questions is 4 hours.

001

The tables below show historical data on traffic counts for a bridge, averaged by day of the week and month. If a daily count, conducted on a Wednesday in April is 19,545, the AAWT (Average Annual Weekday Traffic) for planning purposes is most nearly

A. 16,300

B. 17,250

C. 22,830

D. 23,440

Day of the week	ADT
Sunday	12,760
Monday	18,985
Tuesday	20,765
Wednesday	19,882
Thursday	20,349
Friday	16,889
Saturday	13,725
TOTAL	123,355

Month	ADT
January	17,756
February	16,772
March	19,674
April	21,983
May	20,935
June	16,783
July	15,887
August	16,785
September	19,836
October	19,356
November	20,128
December	19,785
TOTAL	225,680

002

The boundaries of a site form a triangle as shown below. The area of the site (acres) is most nearly

A. 16

B. 24

C. 32

D. 48

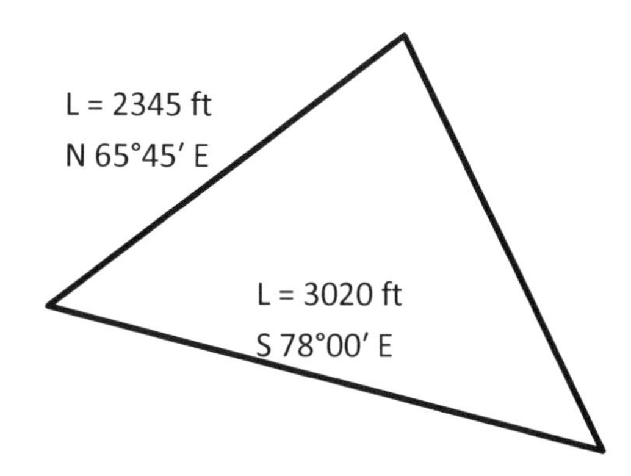

003

A reinforced concrete pipe of 36 inch outer diameter connects two manholes MH-1 and MH-2 as shown in the figure below. At station 13 + 05.10, the ground surface has a low point elevation of 242.35 ft. At this location, the soil cover (feet) is most nearly:

A. 3.95
B. 4.12
C. 4.56
D. 4.72

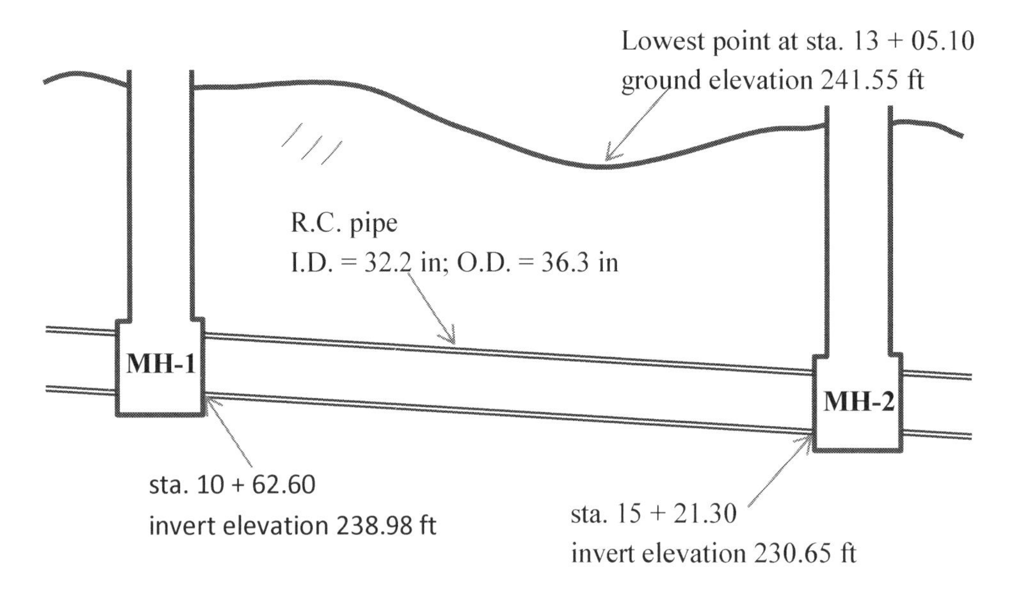

004

A roadside gutter is in the shape of a symmetric v-channel with 3H:1V sideslopes. The gutter is to be lined with a 3 inch thick concrete liner as shown. If the concrete material + placement cost is $232/yd^3, then the cost of constructing the gutters ($/mile) is most nearly

A. 36,000
B. 41,000
C. 72,000
D. 82,000

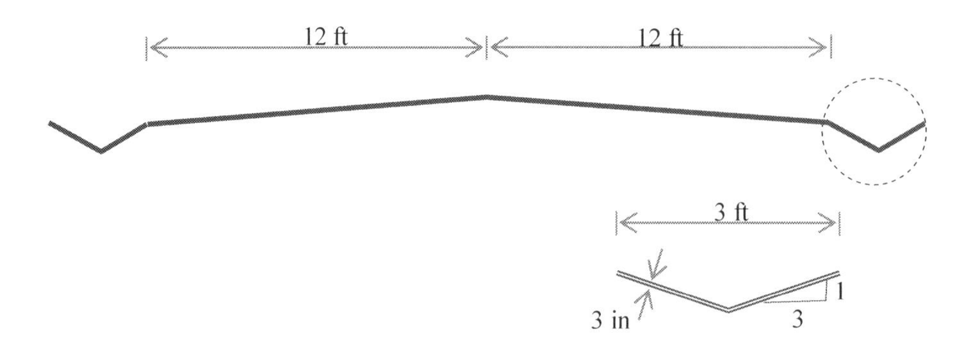

005

The activity on node network for a project is as shown below. All relationships are finish to start unless otherwise indicated. The table on the right shows pertinent data. The early start date (weeks) for activity F is:

A. 15

B. 16

C. 17

D. 18

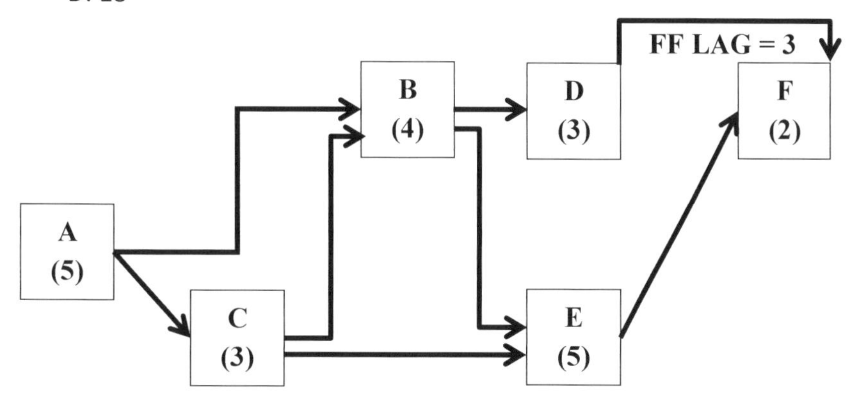

006

A parabolic vertical curve must connect a tangent of slope +5% to another of slope -3% as shown below. The two tangents intersect at a point at station 11 + 45.20 and at elevation 310.56 ft. A sewer (circular CIP) with crown elevation 302.65 ft exists at station 12 + 30.05. If minimum soil cover of 30 inches is required above the sewer pipe, the required length of curve (feet) is most nearly

A. 510

B. 570

C. 635

D. 815

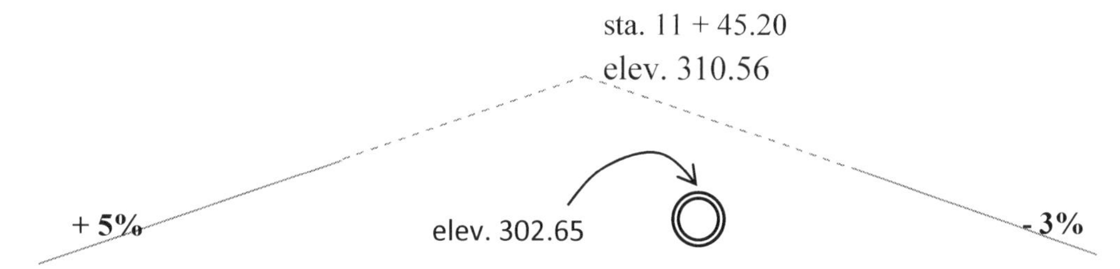

007

A cantilever retaining wall is supported by a 3-foot thick footing as shown. The drains behind the wall become clogged and groundwater rises to the top of the horizontal backfill. The total horizontal earth pressure resultant (lb/ft) acting on the retaining wall is most nearly:

A. 22,300
B. 20,650
C. 18,770
D. 13,300

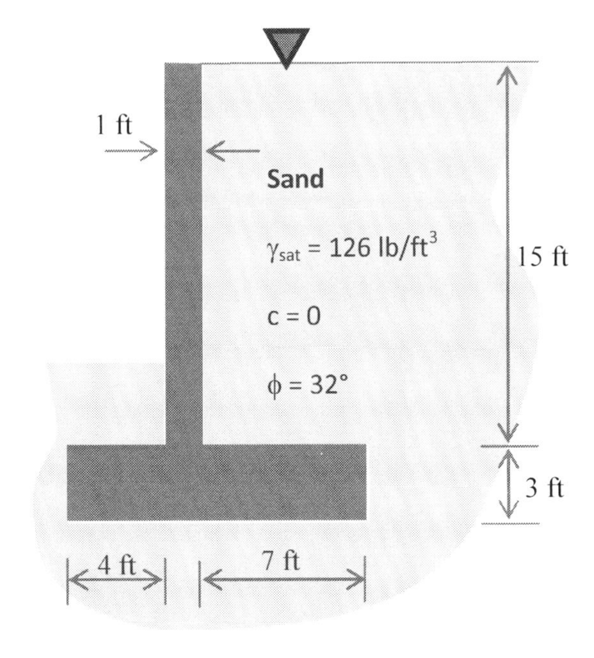

008

A 10-ft layer of varved clay is overlain by a 12 ft thick sand layer, as shown. The water table is originally at a depth of 5 ft below the ground surface. Prior to construction, the water table is lowered by 7 ft, to the bottom of the sand layer. Three months after lowering the water table, settlement (inches) due to consolidation of the clay layer is most nearly

A. 0.9
B. 1.9
C. 2.6
D. 4.0

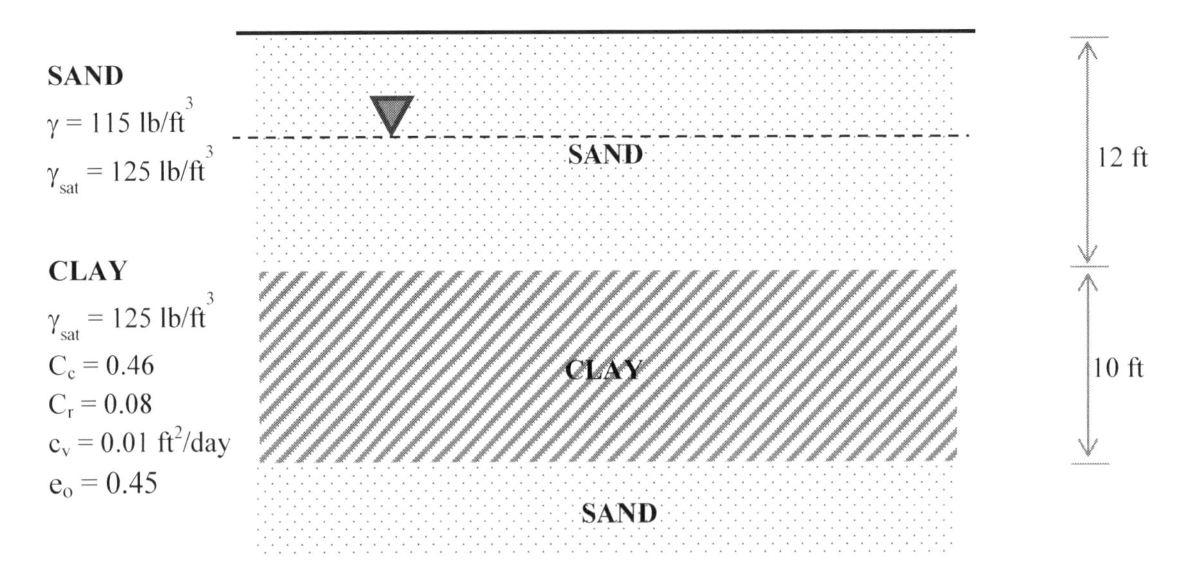

009

Identify the correct shape of the bending moment diagram for the beam loaded as shown below.

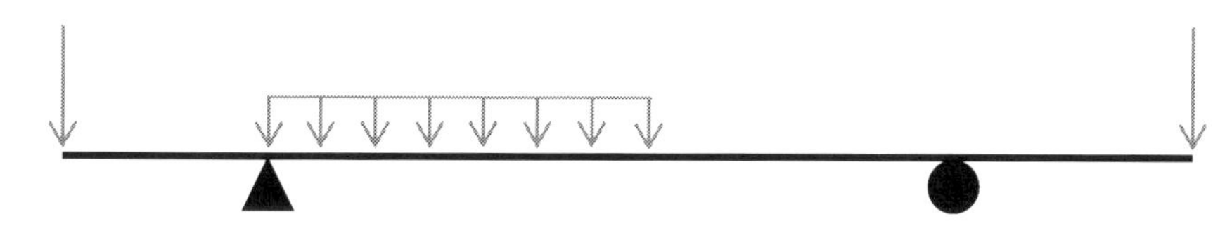

A.

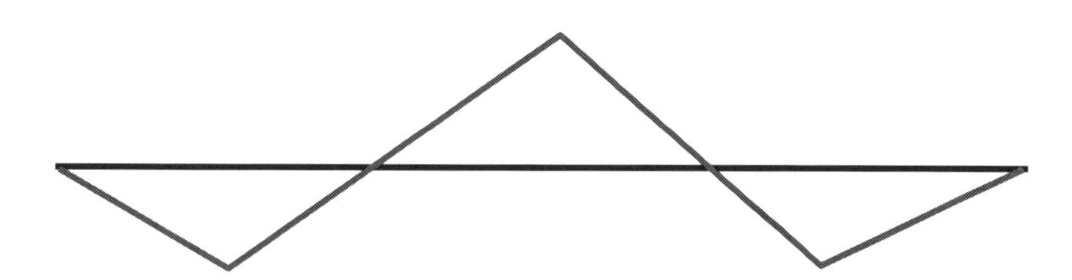

B.

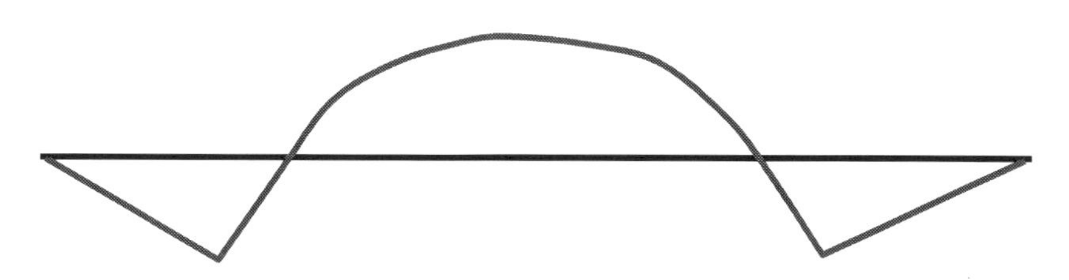

C.

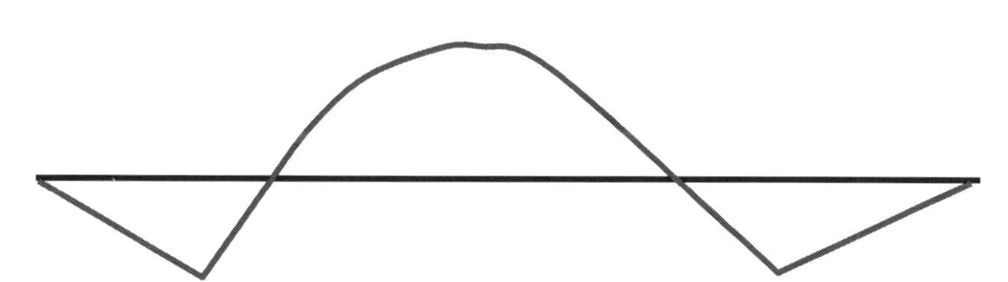

D.

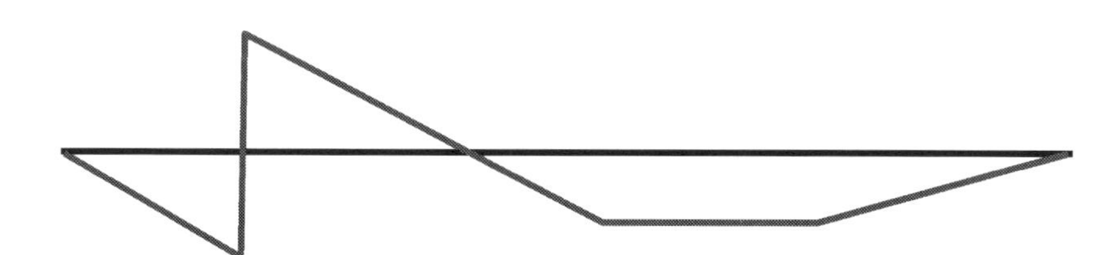

010

A rectangular post is subject to an eccentric load P as shown. The maximum compressive stress (MPa) is most nearly

A. 1.0
B. 1.6
C. 2.6
D. 4.0

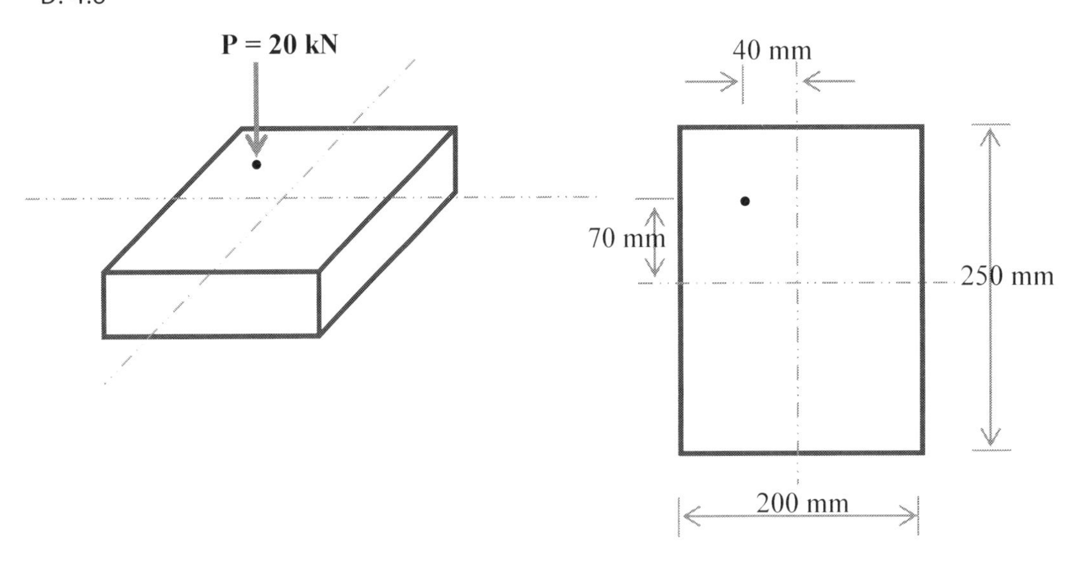

011

A trapezoidal open channel conveys flow at a uniform depth of 5 ft as shown below. The Manning's n = 0.015. The bottom width of the channel = 20 ft and longitudinal slope of the channel floor is 0.8%.

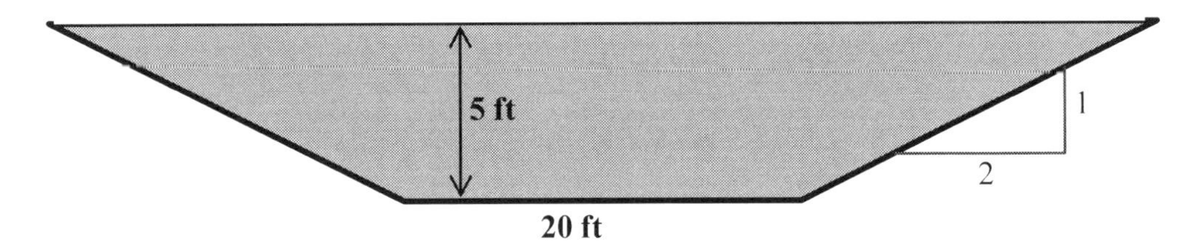

The flow rate (MGD) conveyed by the channel is most nearly:

A. 500
B. 1000
C. 2000
D. 3000

012

A 5 ft x 5 ft square footing transfers a column load of 140 kips to a sandy soil as shown. The depth of the footing is 3 ft. The factor of safety against general bearing capacity failure is most nearly

A. 3.3

B. 2.7

C. 2.0

D. 1.3

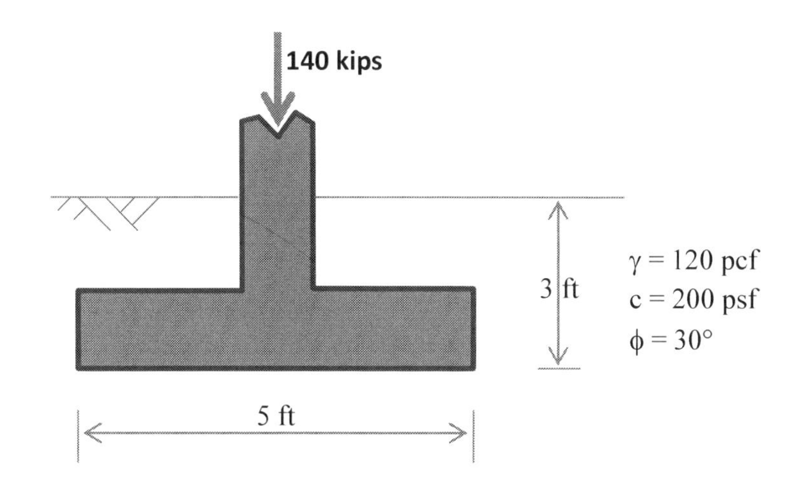

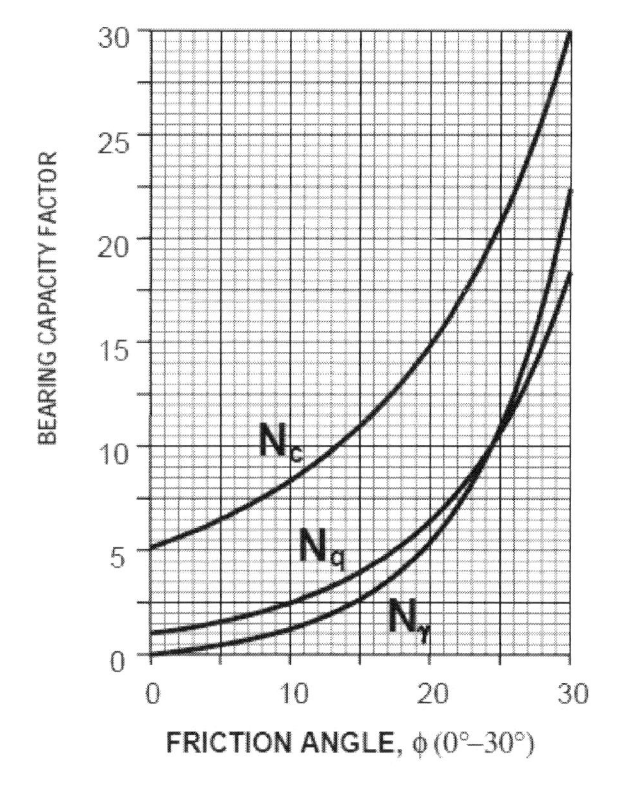

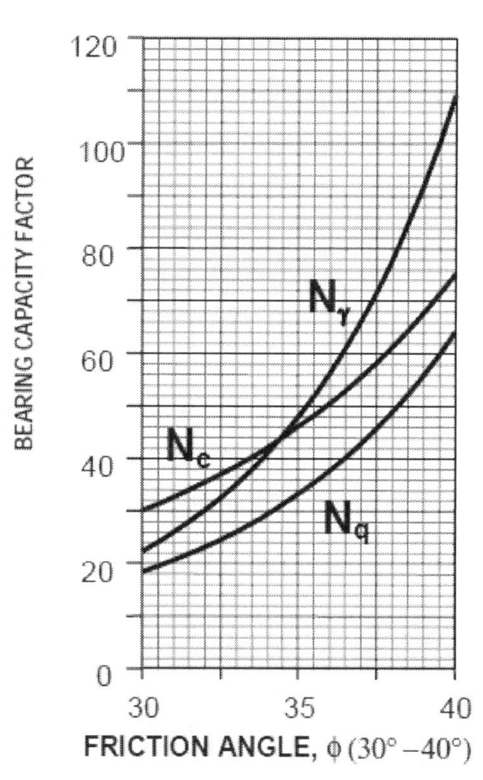

013

A wall panel is composed of plywood sheathing supported by 2x6 studs (nominal dimensions 1.5 in x 5.5 in) spaced every 28 inches as shown. The studs are supported by longitudinal members every 10 ft. The wall experiences a normal wind pressure of 30 psf. The maximum bending stress (lb/in^2) in the studs is most nearly:

A. 800
B. 1000
C. 1200
D. 1400

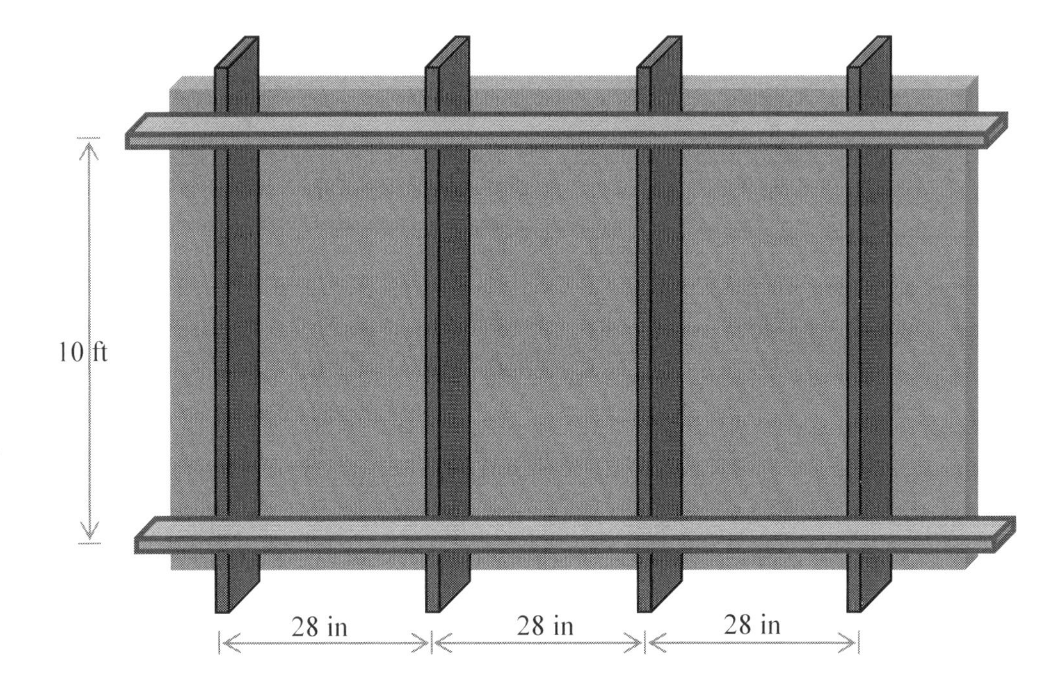

014

The table below shows cross section areas of cut and fill recorded at 5 stations spaced at 100 ft.

Station	Area (ft^2) CUT	FILL
0 + 0.00	245.0	123.5
1 + 0.00	312.5	76.3
2 + 0.00	411.5	0.0
3 + 0.00	234.5	88.4
4 + 0.00	546.2	214.5

The net earthwork volume (yd^3) between stations 0 + 0.00 and 4 + 0.00 is most nearly

A. 3640 (cut)
B. 3780 (cut)
C. 3640 (fill)
D. 3780 (fill)

015

A site needs soil compacted to 90% of the Proctor maximum dry density. The results of the Proctor test are shown below.

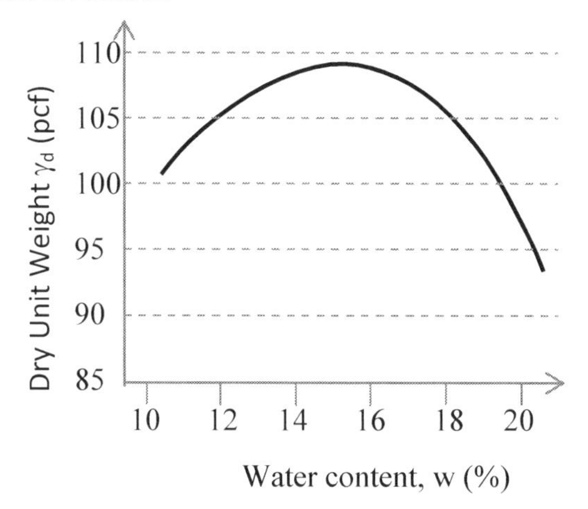

The volume of the embankment is 50,000 ft^3. If borrow soil is available at γ = 120 pcf and moisture content = 14%, the volume of soil (yd^3) needed from the borrow pit is most nearly:

A. 1,400
B. 1,730
C. 2,080
D. 2,800

016

Isohyets showing precipitation depth are shown in the figure. The accompanying table shows total area enclosed by each close contour. The average precipitation depth (inches) is most nearly:

A. 0.46
B. 0.51
C. 0.58
D. 0.62

Contour	Area enclosed (acres)	Precip. depth (inches)
1	268	0.2
2	121	0.6
3	45	0.9

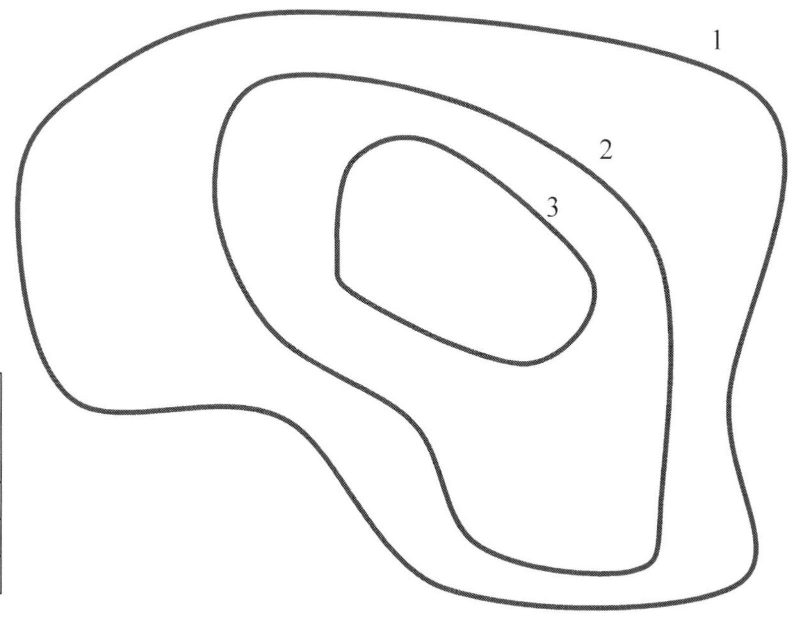

017

A 24 inch diameter reinforced concrete pipe (C = 100) conveys water a flow rate = 12.5 ft^3/sec. The head loss due to friction (feet per mile) is most nearly:

A. 11.5
B. 14.6
C. 16.3
D. 18.3

018

Runoff flow from a development is held in a detention pond until it is 78% full, at which point, it empties through a weir. The capacity of the pond is 760,000 gallons. If after a rainfall event, the average rate of inflow into the pond occurs at 2 ft^3/sec, the length of time (hours) before the pond starts to empty is most nearly:

A. 7.5
B. 9.3
C. 11.0
D. 13.7

019

For the truss shown below, the axial force (kips) in member CD is most nearly:

A. 125 (tension)
B. 125 (compression)
C. 150 (tension)
D. 150 (compression)

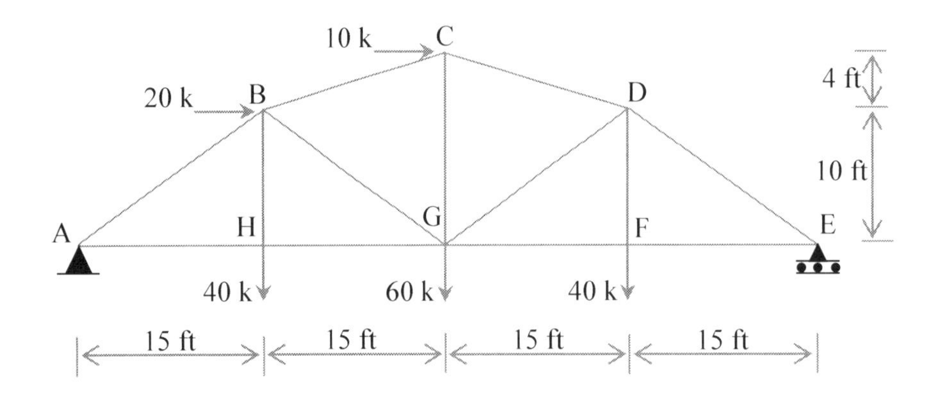

020

A 6-inch thick riprap layer is used as protection for the earth slope (θ = 30°) shown below. The factor of safety for slope stability is most nearly:

A. 1.25

B. 1.35

C. 1.45

D. 1.55

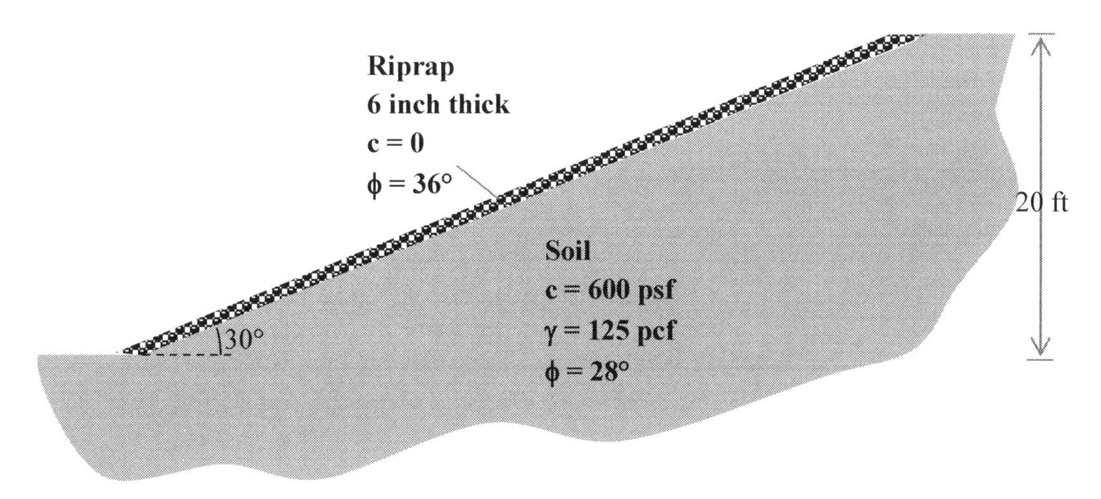

021

Particle size distribution of a soil sample is summarized in the table below. The fines were tested using Atterberg apparatus to obtain the following results:

Liquid limit 45

Plastic limit 21

Sieve size	% finer
1.0 inch (25.4 mm)	100
0.5 inch (12.7 mm)	92
No. 4 (4.75 mm)	75
No. 10 (2.00 mm)	62
No. 40 (0.425 mm)	52
No. 100 (0.15 mm)	45
No. 200 (0.075 mm)	28

The classification of the soil according to the USCS is:

A. GW

B. SM

C. SC

D. GC

022

A soil sample yields the following results:
Mass of wet soil = 1685 g
Volume of wet soil = 855 cc
Mass of soil after oven drying = 1418 g
If the specific gravity of soil solids is taken as 2.65, the void ratio is most nearly:

A. 0.3
B. 0.4
C. 0.5
D. 0.6

023

2x6 stud columns (actual dimensions 1.5 in x 5.5 in) are connected to plywood sheathing as shown. The modulus of elasticity of timber is $E = 1.5\text{x}10^6$ lb/in^2. The Euler buckling load (kips) for each column is most nearly:

A. 9
B. 15
C. 21
D. 26

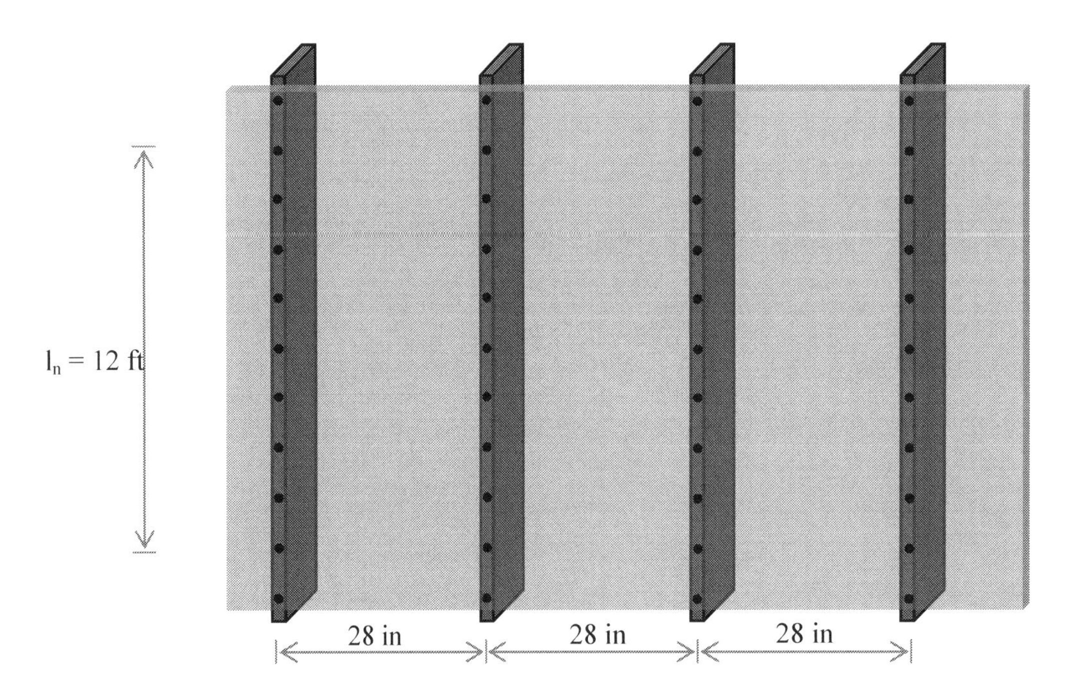

024

Water stored in a large reservoir (surface elevation 324.5 ft above sea level) empties through a 24 inch diameter pipe as shown. The far end of the pipe is at elevation 295.8 ft above sea level. The discharge (ft^3/sec) through the pipe is most nearly:

A. 50
B. 38
C. 32
D. 22

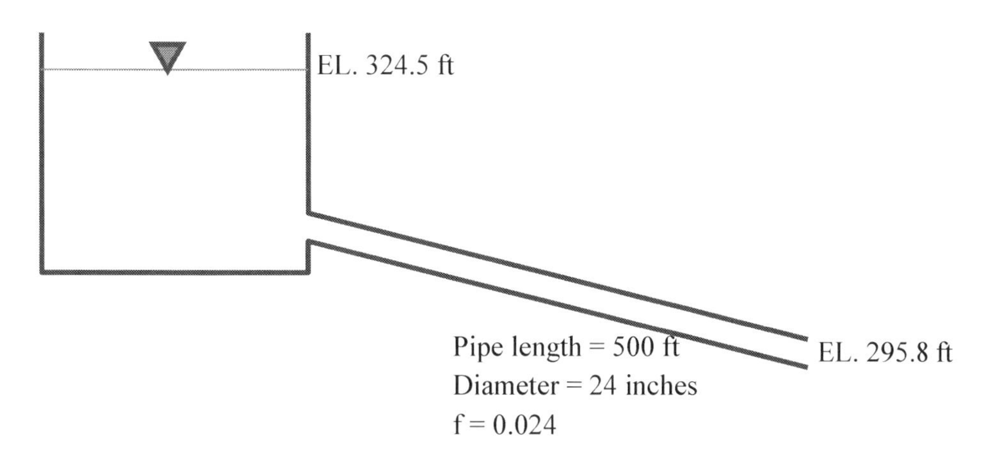

025

Surface runoff from a catchment area has the longest hydraulic path as shown in the figure below. Time for sheet flow, t_s = 5 minutes and for ditch flow t_d = 13 minutes. A collector pipe, in which flow occurs at an average velocity of 5 feet/sec, of length 1200 ft has an inlet at point A and discharges into a main sewer at point B. Intensity-duration-frequency curves are obtained from historical precipitation data. The design intensity (in/hr) to be used for the design of the sewer mains for a 20 year storm is most nearly:

A. 1.5
B. 1.8
C. 2.1
D. 2.4

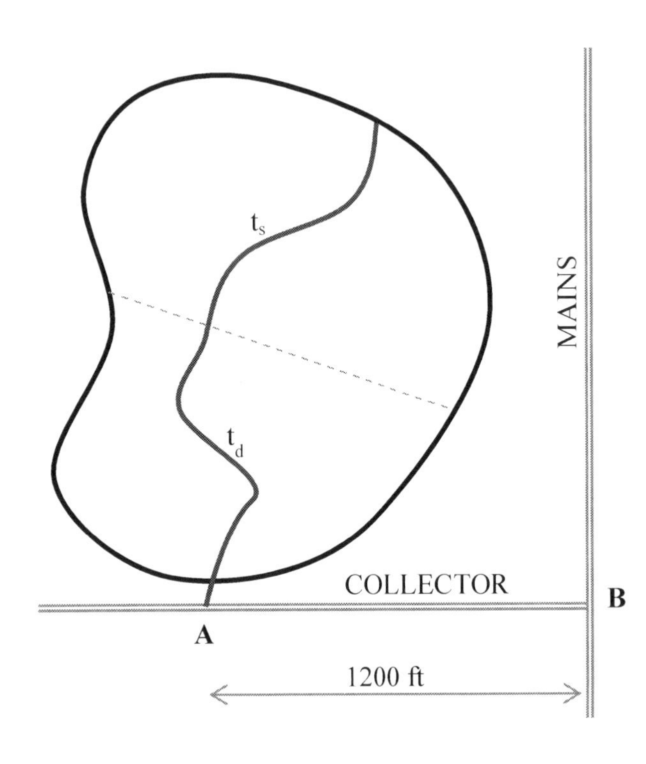

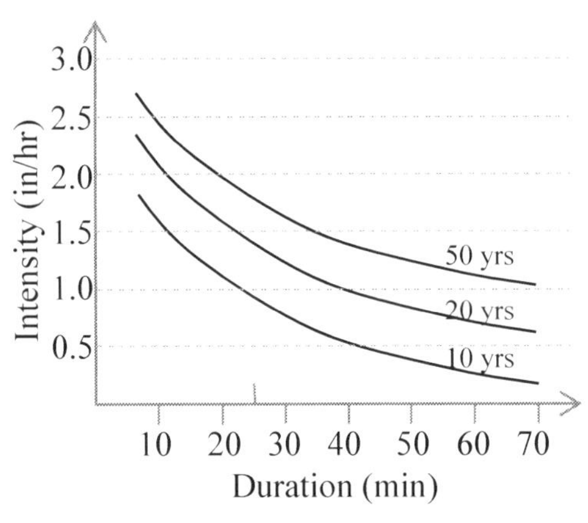

026

A circular horizontal curve has the coordinates (feet) of the PC, PI and PT as shown on the figure.

The degree of curve (degrees) is most nearly:

A. 10
B. 13
C. 16
D. 19

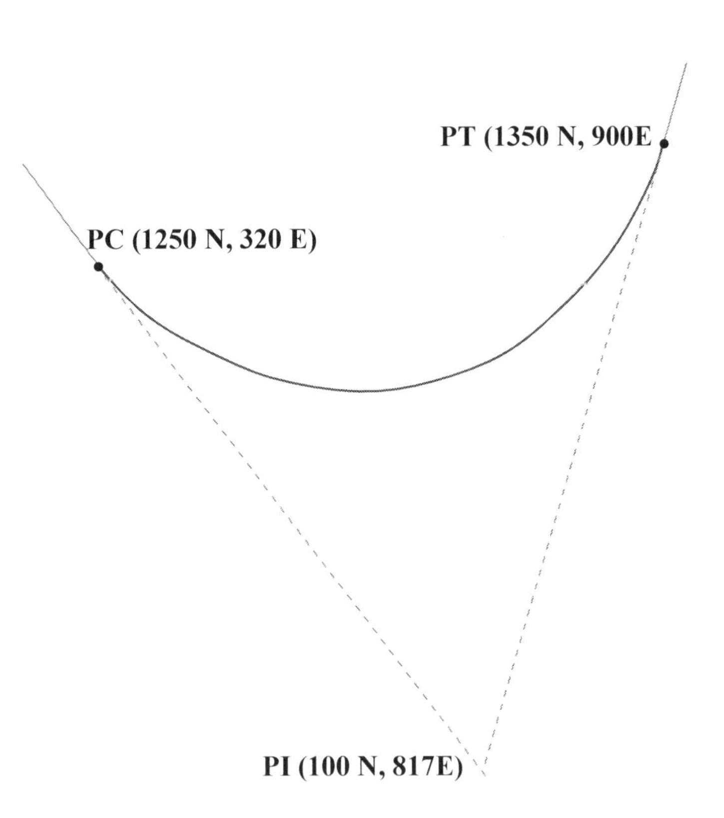

027

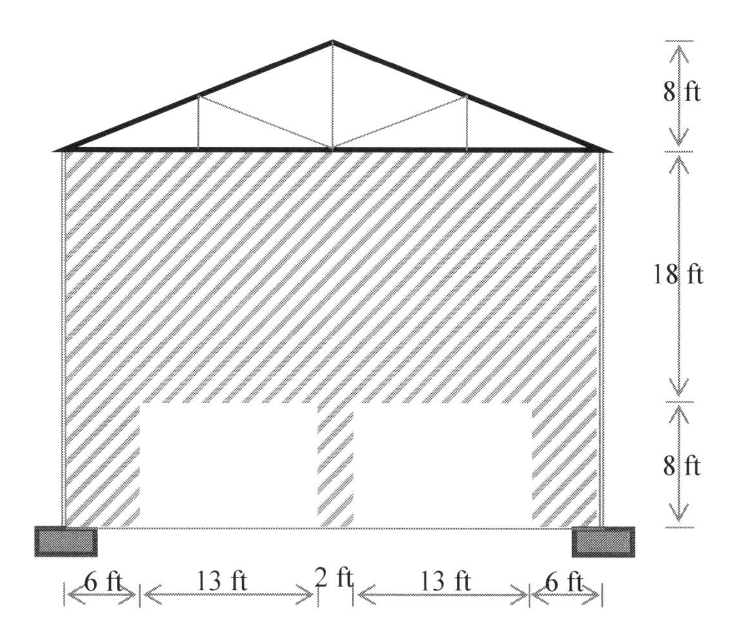

A 110 ft long warehouse shed has metal sheathing forming the sides (no openings) and front and back (with two door openings on each surface) and the roof. The total surface area of metal sheathing (sq. ft.) is most nearly:

A. 12,652
B. 12,456
C. 12,331
D. 12,123

028

A crane with a 40 ft boom is used to lift a 4 ton load as shown. The total weight of the crane and ballast is 4.5 tons acting at the effective location indicated as CG on the figure. The weight of the boom is 800 lb. Each outrigger leg is supported by a circular pad with a diameter = 3 feet

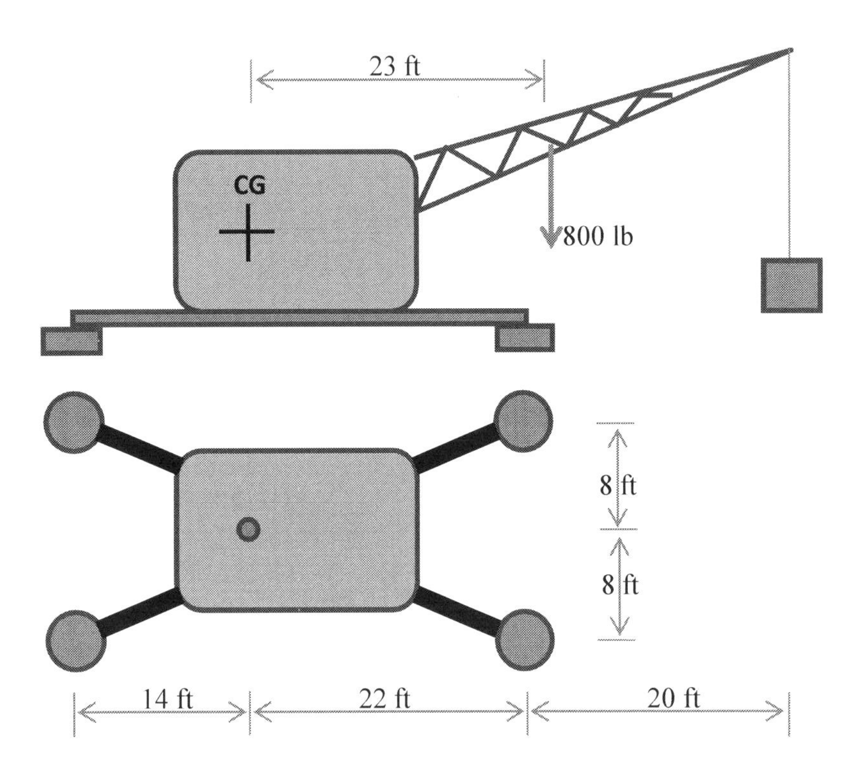

The maximum soil pressure (lb/ft^2) under the outrigger pads is most nearly:

A. 1,200
B. 1,600
C. 2,000
D. 2,400

029

Which of the following statements is true about deflection of concrete beams?

I. The deflection is calculated using the moment of inertia of the uncracked section
II. The deflection is calculated using a moment of inertia equal to half the gross moment of inertia
III. The deflection is calculated using the moment of inertia of the cracked section
IV. The deflection is calculated using a moment of inertia less than the gross moment of inertia

A. I and IV
B. IV only
C. II only
D. II and III

030

The figure shows a stress-strain diagram based on a tension test of a steel test coupon.

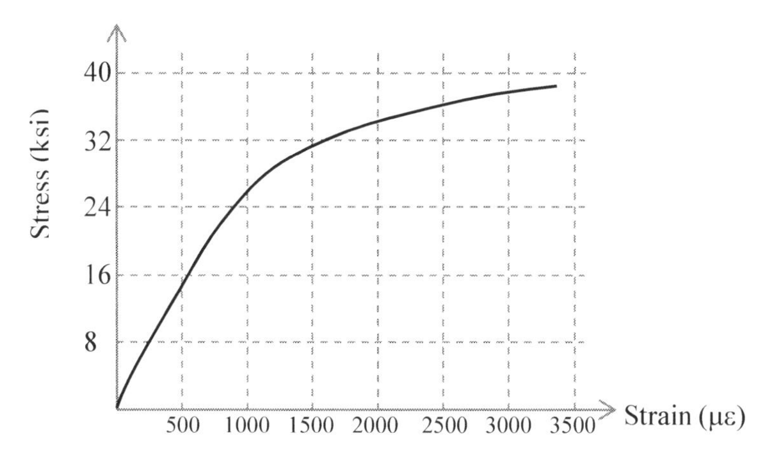

The yield stress (kip/in^2) is most nearly

A. 26
B. 30
C. 34
D. 38

031

A layer of coarse sand (thickness = 15 feet) supports a mat foundation that exerts a net uniform pressure of 600 psf at a depth of 3 feet below the surface as shown. The compression of the sand layer (inches) is most nearly:

A. 0.1
B. 0.2
C. 0.3
D. 0.4

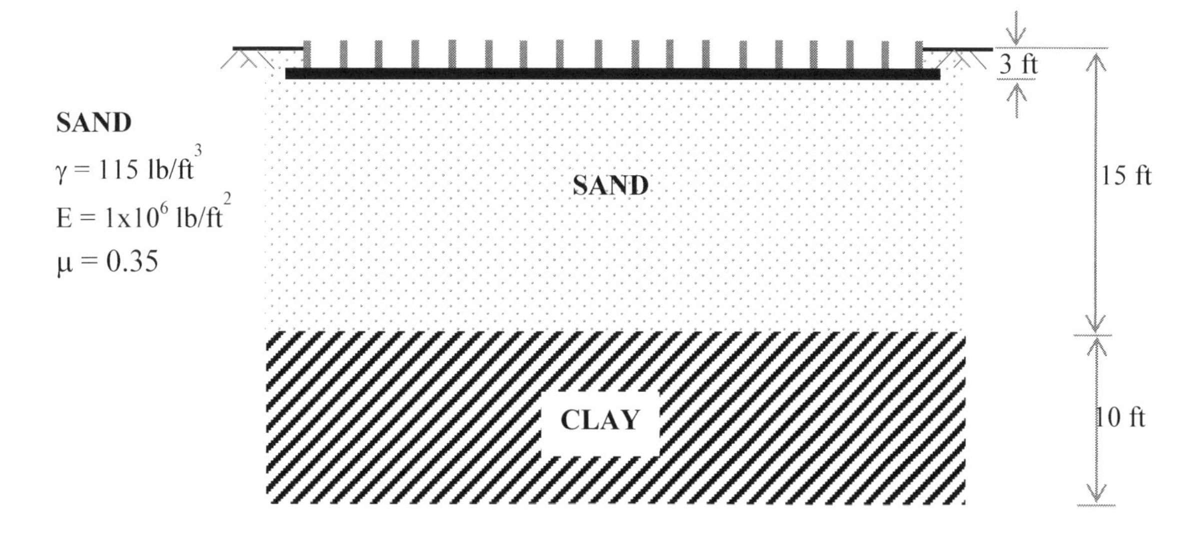

032

Which of the following statements about total float (TF) in a CPM network is not true? Use the following nomenclature – ES (early start), EF (late finish), LS (late start), LF (late finish), D (duration).

I. TF = LF – ES – D

II. TF = LS – EF – D

III. TF = LS – EF + D

IV. TF = LS – ES

A. I

B. II

C. III

D. IV

033

Which of the following statements is/are true?

I. For long term stability analysis of clay slopes, results of the CD triaxial test must be used.

II. The CD triaxial test takes longer to perform than the UU test.

III. The UU triaxial test takes longer to perform than the CD test

IV. Pore pressure measurements must be made during the CD test

A. I, II only

B. II, III only

C. III only

D. all of them

034

A test strip shows that a steel-wheeler roller, operating at 3 mph, can compact a 0.5 ft. layer of material to a proper density in four passes. The width of the drum is 8.0 ft. The roller operates 50 min per hour. The number of rollers required to keep up with a material delivery rate of 540 bank cubic yards/hr is most nearly: (1 bank cubic yard = 0.83 compacted cubic yard):

A. 4

B. 3

C. 2

D. 1

035

A simply supported steel beam carries a single concentrated load at midspan as shown. The beam as the following properties: Area A = 29.4 in^2; I_x = 1490 in^4; I_y = 186 in^4; Z_x = 198 in^3; Z_y = 54.9 in^3; S_x = 175 in^3; S_y 35.7 = in^3.

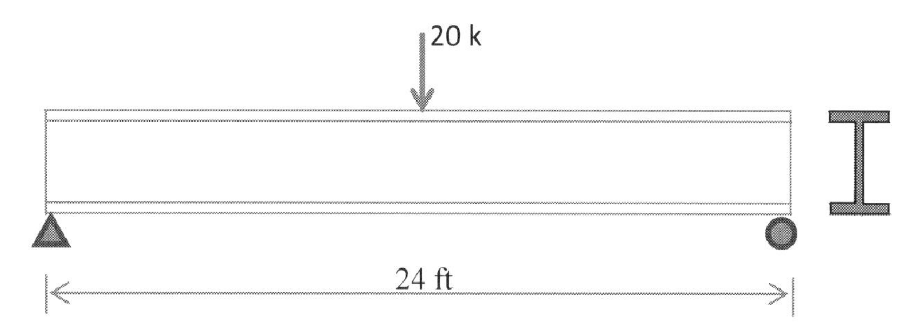

The maximum deflection (inches) is most nearly:

A. 0.25

B. 0.50

C. 0.75

D. 0.90

036

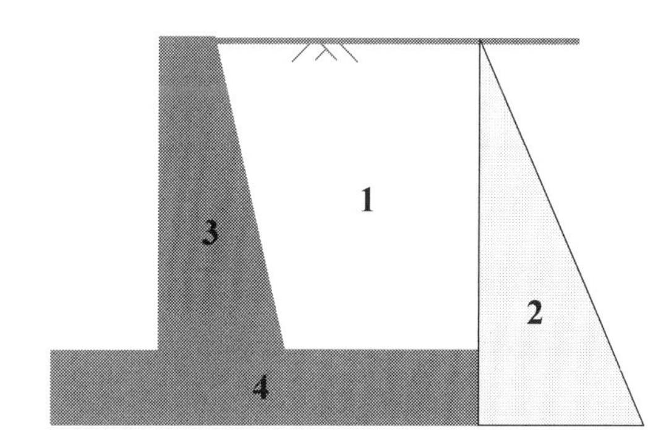

	Component	Resultant Force
1	Backfill soil	12,600 lb/ft
2	Active earth pressure	5,400 lb/ft
3	Concrete wall stem	3,500 lb/ft
4	Concrete wall footing	4,500 lb/ft

A concrete retaining wall has a level backfill behind it. The coefficient of friction between the wall footing and the soil is 0.6. The table to the right summarizes the forces acting on the retaining wall. The factor of safety against sliding is most nearly:

A. 0.9

B. 1.6

C. 2.3

D. 2.9

037

The 'first flush' runoff depth (assumed to be 1 inch) from a 120 acre watershed collects in a 2 acre detention pond. The sediment load carried by the runoff is 5 g/L. Bulk density of sediment is 80 lb/ft^3. The loss of depth (inches) per rainfall event is most nearly:

A. 0.125

B. 0.25

C. 0.50

D. 0.75

038

Several cylindrical steel samples are tested in a Universal Testing Machine to obtain the results obtained below:

Sample	Diameter (in)	Sample length (in)	Breaking load (lb)	Elongation at Failure (in)
1	0.504	5.66	9050	0.412
2	0.498	7.34	8865	0.516
3	0.509	7.55	9235	0.543
4	0.503	8.12	9110	0.581
5	0.512	7.12	9565	0.505

The average breaking strain ($\mu\varepsilon$) is most nearly:

A. 38,500

B. 51,200

C. 71,500

D. 78,300

039

A circular conduit of diameter 48 inches conveys water at a depth of 30 inches as shown below. The interior of the concrete pipe is coated to yield n = 0.014 (assumed constant with varying depth).

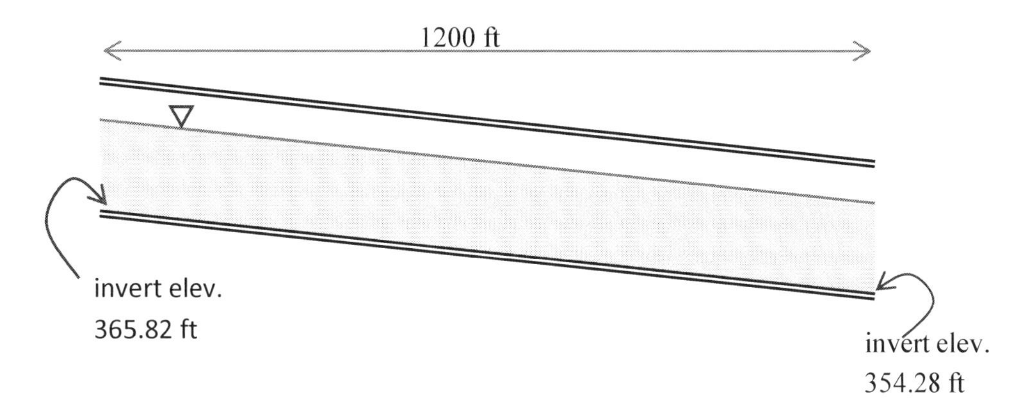

The velocity (ft/sec) is most nearly:

A. 8.2
B. 9.3
C. 10.4
D. 11.3

040

Which of the following techniques are commonly used for construction adjacent to historic structures?

I. Underpinning
II. Anchor rod and deadman
III. Slurry walls
IV. Compaction piles

A. I and II
B. I and III
C. II and III
D. II and IV

THIS IS THE END OF THE BREADTH EXAM

CONSTRUCTION DEPTH EXAM

FOR THE

CIVIL PE EXAM

The following set of 40 questions (numbered 201 to 240) is representative of a 4-hour depth (PM) exam for CONSTRUCTION according to the syllabus and guidelines for the Principles & Practice (P&P) of Civil Engineering Examination (updated January 2015) administered by the National Council of Examiners for Engineering and Surveying (NCEES).

The time allocated for this set of questions is 4 hours.

101

The mass diagram shown below represents the earthwork for a highway project. Stations are marked every 100 feet. The number of hauling zones is equal to

A. 4
B. 3
C. 2
D. 1

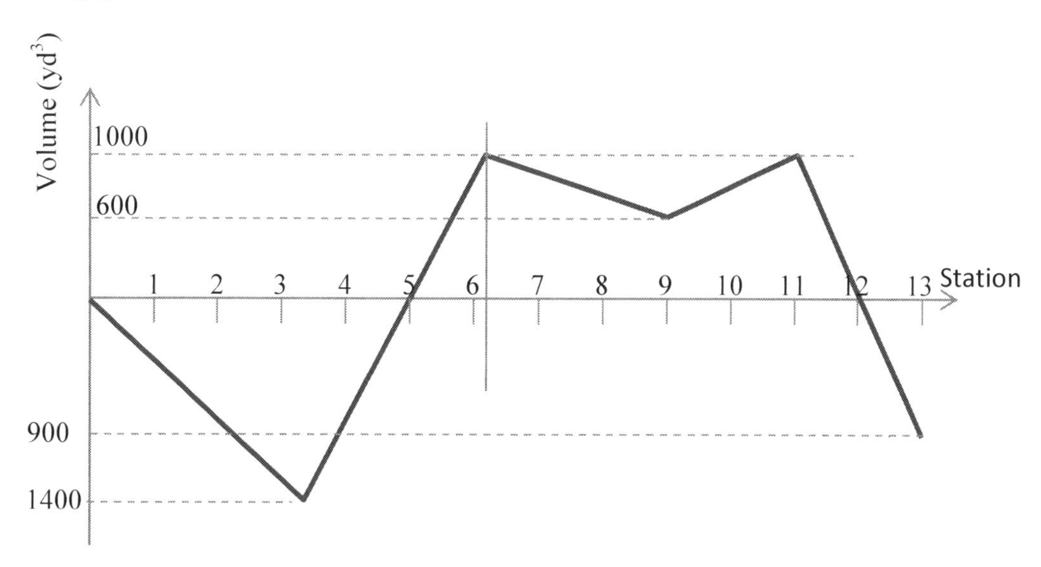

102

The plan of an excavation is shown in the figure below. Each grid square is 100 ft x 100 ft. Numbers at each vertex represent depth of cut in feet. The total volume of excavation (yd^3) is most nearly:

A. 23,000
B. 46,000
C. 69,000
D. 92,000

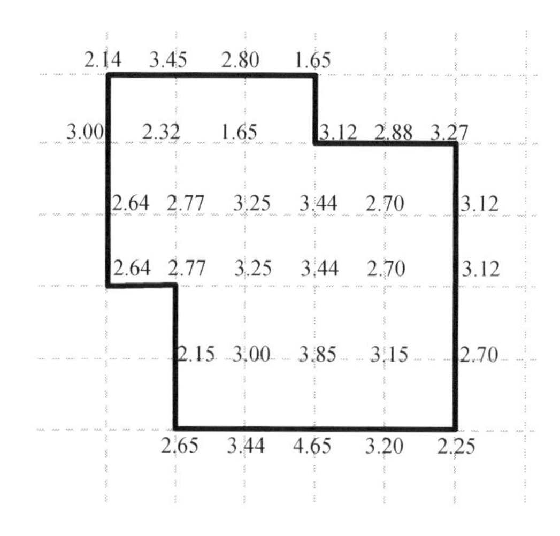

103

An earth-mover has a capacity of 2.4 yd^3 (loose soil). The cycle time = 4 min 30 seconds. If the embankment soil has a bank density = 124 lb/ft^3 and a 15% swell potential, the production (yd^3 bank measure per hour) is most nearly:

A. 20
B. 24
C. 28
D. 32

104

The WSDOT pile driving formula for the ultimate capacity R_u (kips) is given by

$$R_u = 6.6 F_{eff} W H \ln(10N)$$

where

F_{eff} = hammer efficiency,
W = weight of the ram (kips)
H = vertical drop of the hammer (ft)
N = average penetration resistance (blows per inch) for the last 4 inches of driving.

A pile is driven into dense clay by a hammer with efficiency 70%. The ram weighs 800 lb and the fall height is 8 ft. If the last 4 inches of pile driving requires 7 blows, the developed ultimate bearing capacity for the pile is most nearly:

A. 54
B. 75
C. 96
D. 120

105

The figure below shows the activity on node network for the project. Numbers adjacent to the activity labels are duration in weeks.

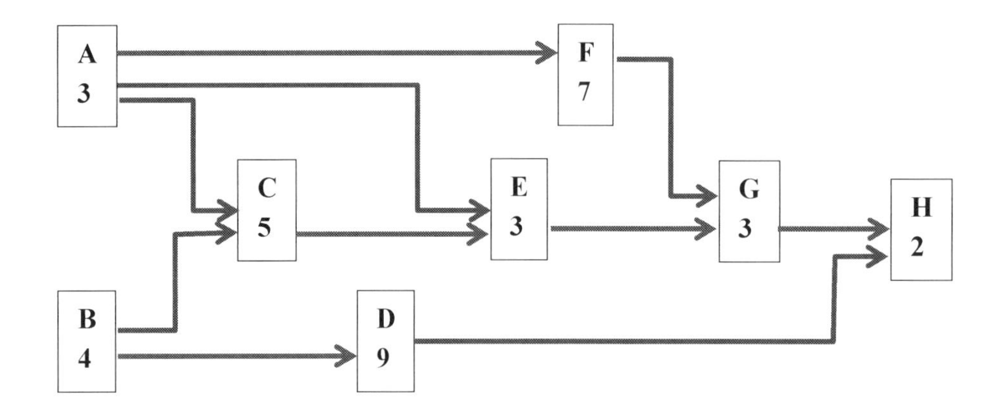

The early start time (week) for activity G is:

A. 12
B. 13
C. 14
D. 15

106

Which of the following weld symbols is/are incorrect?

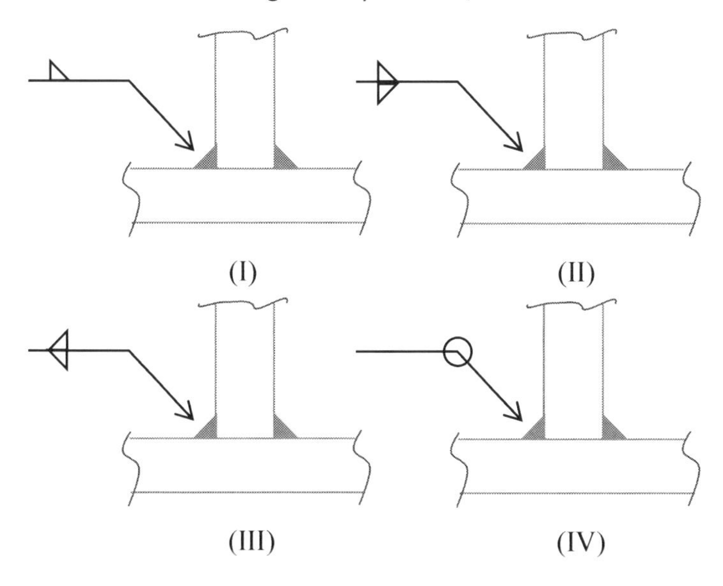

A. I and IV only
B. I and II only
C. II and III only
D. I, III and IV only

107

A concrete mix is prepared by using the following proportions (by weight) 1:1.7:2.5.

SG of cement = 3.15
SG of SSD sand = 2.60
SG of SSG coarse aggregate = 2.70
Air content = 3%
Water cement ratio = 0.40

The specific gravity of the concrete mix is most nearly

A. 2.44
B. 2.41
C. 2.36
D. 2.32

108

Which of the following is an ASTM approved method for predicting compression strength of concrete?

A. PCA empirical method
B. Datum temperature method
C. Arrhenius method
D. Pullout strength measurement

109

Fresh concrete made with Type II cement (unit weight = 130 lb/ft^3) is poured into a column form at a rate of 8 feet/hour. The cement contains no retarders. The height of the column is 14 feet. Concrete temperature = 70°F. The lateral pressure (lb/ft^2) at the midheight of the column is most nearly:

A. 865
B. 910
C. 1035
D. 1090

110

A multistory concrete building is being constructed using one level of shoring and two levels of reshores. The dead load of a floor slab is designated D. Construction live load (due to presence of workers and equipment) is 0.3D, weight of forms and shores is 0.1D and weight of reshores is 0.05D.

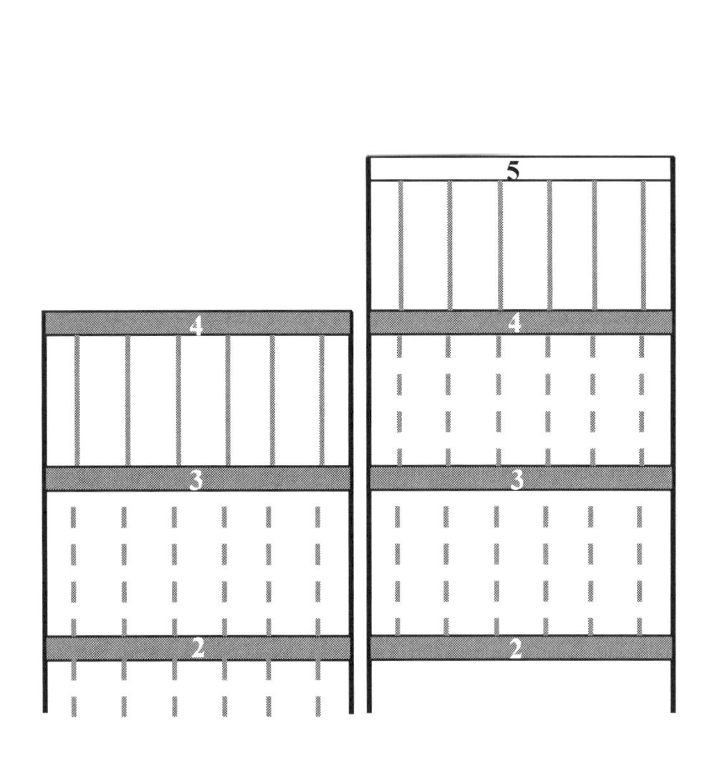

Shore Loads Before Step	Loads carried by Floors			Shore Loads After Step
	Before step	Change	After Step	
	0			
	1.0D			
0				
	1.0D			
0				
	1.0D			
0				

At a certain stage of the project, where the last poured slab (no. 4) has hardened and is self-supporting. During the next stage of construction, the reshores below level 2 are removed and flown up to level 4 and the slab for floor 5 is poured. At the end of this stage, the load in the reshores below level 4 is:

A. 1.34D

B. 1.12D

C. 1.05D

D. 0.98D

111

Cantilevered soldier piles spaced at 8 ft provide support to 3 inch thick lagging which retains sand to a depth of 16 feet as shown. The uniform horizontal pressure exerted by the sand is 470 lb/ft^2. The uniform load (lb/ft) acting on each soldier pile is most nearly:

A. 3750

B. 5600

C. 7500

D. 9000

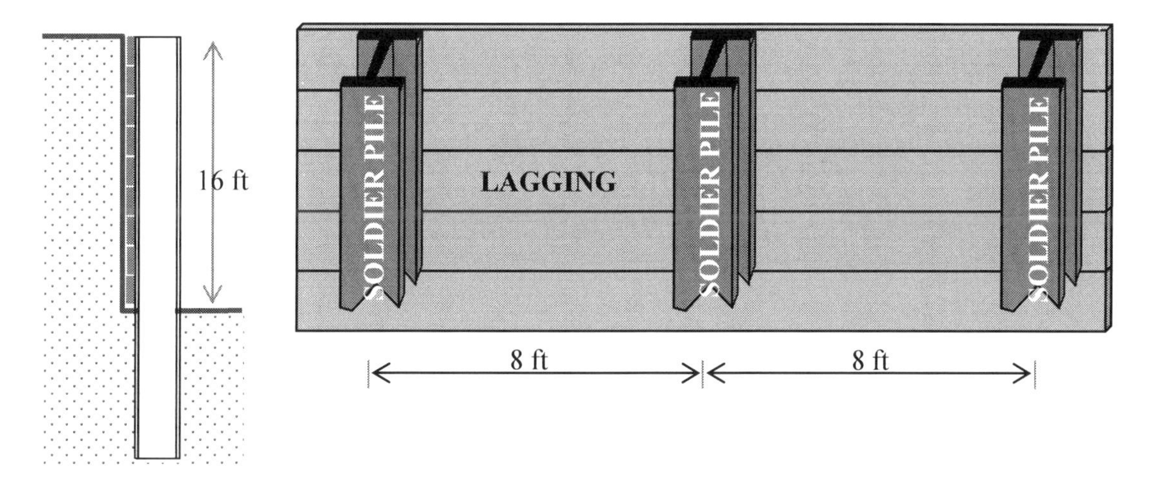

112

According the guidelines of the OSHA 1926 regulations (Title 29 CFR), the minimum light intensity (foot-candles) for carpenter shops located on a construction site is most nearly:

A. 3

B. 5

C. 10

D. 30

113

The mass diagram shown below represents the earthwork for a highway project. Stations are marked every 100 feet. Which of the following statements is/are true?

I. Earthwork is balanced between stations 0 + 00 and 12 + 00

II. Station 5 + 00 represents a transition from a fill to a cut

III. Station 9 + 00 represents a transition from a fill to a cut

A. All of them

B. I and II only

C. II and III only

D. I and III only

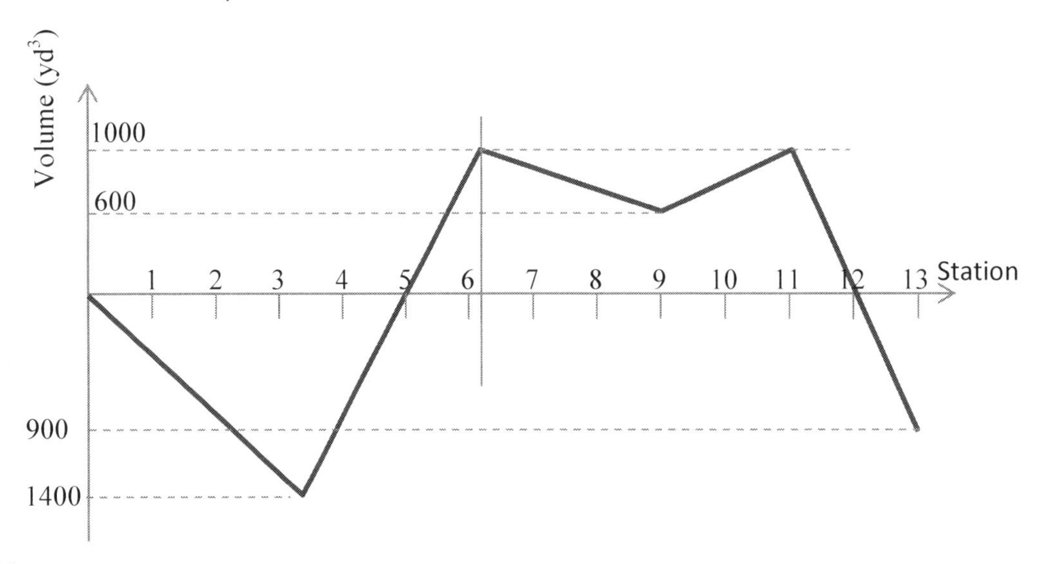

114

Which of the following statements is false?

A. Ground penetrating radar works better for clay soils than sands

B. Seismic reflection methods are little used for shallow ground investigation

C. The primary limitation of conducting a gravity survey is high equipment cost

D. Resistivity surveys are not very effective in site investigation

115

A planned excavation is to be 15 feet deep. The soil will be subject to vibration from pile driving operations. The unconfined compressive strength of the soil is 3250 psf. According the guidelines of the OSHA 1926 regulations (Title 29 CFR), the classification of the soil is:

A. Type A

B. Type B

C. Type C

D. Type D

116

The layout of an excavation plan is shown below. An access road of length 3,800 feet leads from the site entrance to the loading zone in the north-east corner of the site. An excavator (capacity 2.44 yd^3; excavation time = 2 min) digs and transports material from the pit (outlined with a dashed line) to the loading zone. The material is then loaded onto trucks (capacity 50 ft^3) which transport the material offsite using the access road. The mean distance from the excavation pit to the loading zone is 500 feet. The excavator's mean travel speed to the loading zone is 10 mph and the trucks operate at an average speed of 15 mph. The loading time for each truck is 30 seconds.

The minimum number of trucks needed to meet the production rate of the excavator is most nearly:

A. 1
B. 2
C. 3
D. 4

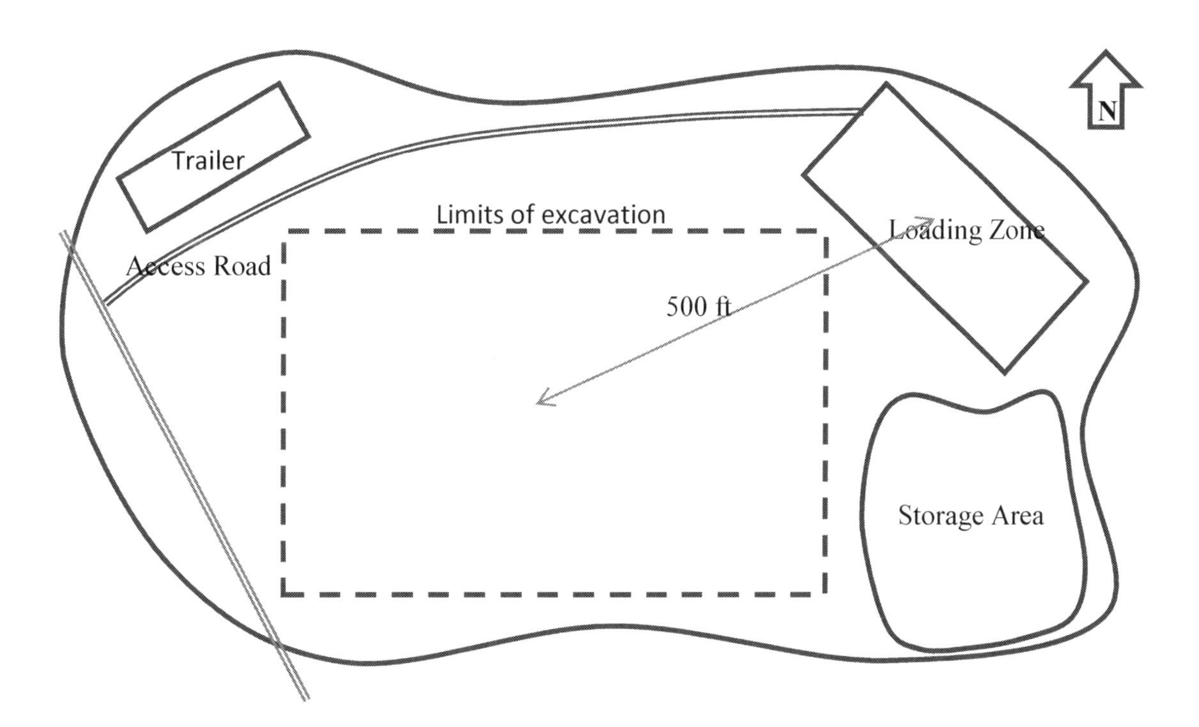

117

The reinforcement pattern for a 6 inch thick slab panel is shown below. All reinforcement indicated is identically placed in a top and a bottom layer. Clear cover to reinforcement is 0.75 inch. Each of the edge strips is reinforced with 3 no. 5 bars (top and bottom, as indicated)

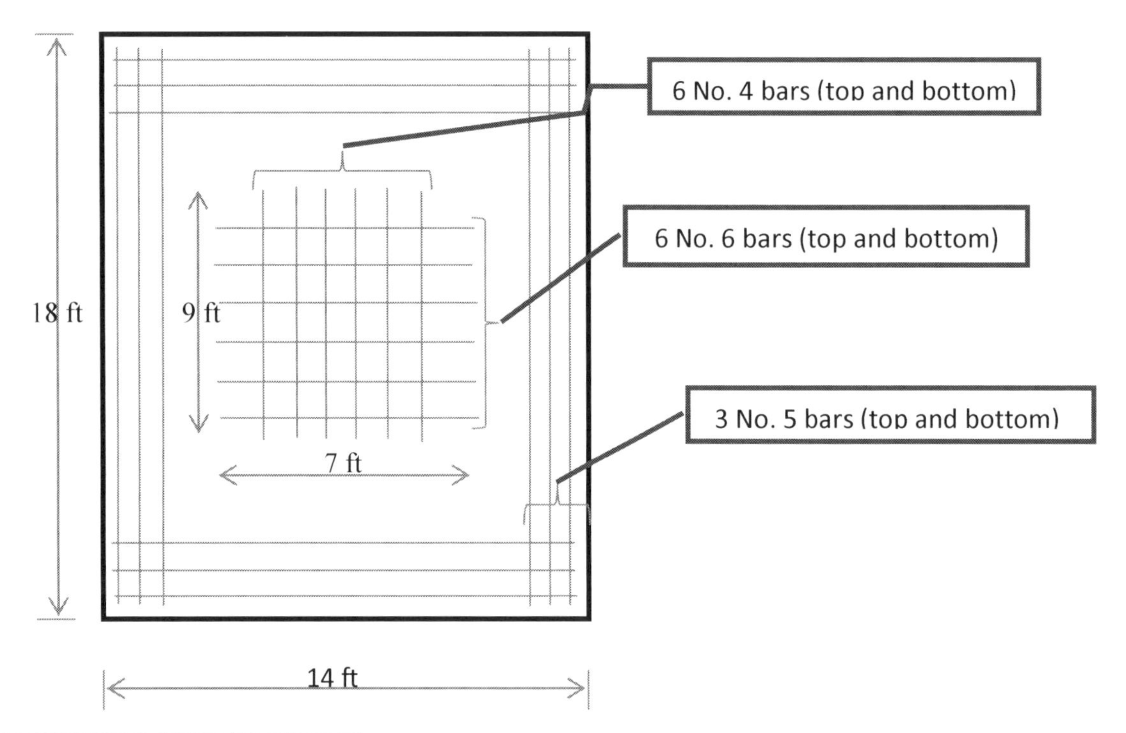

Bar size	Area (in^2)	Weight (lb/ft)
3	0.11	0.38
4	0.20	0.67
5	0.31	1.04
6	0.44	1.50
7	0.60	2.05
8	0.79	2.67

The total weight of reinforcement steel (lb) per slab panel is most nearly:

A. 600
B. 330
C. 300
D. 165

118

A crane is used to lift and place a load on the roof of a building as shown. The length of the boom is 110 feet. Safety regulations do not permit the boom of the crane to come any closer than point A (offset 6 feet vertically and horizontally from the corner of the building).

The maximum distance X (feet) from the front edge of the building that the load can be placed on the roof is most nearly:

A. 30
B. 25
C. 20
D. 15

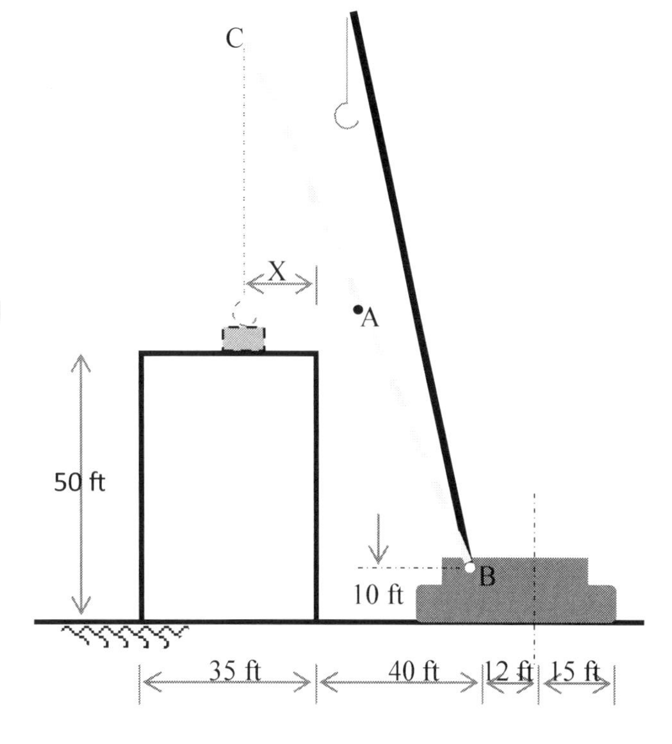

119

An activity chain in a construction project consists of activities A and B. Stochastic estimates for completion times are as follows:

Activity A: Optimistic duration estimate = 4 weeks
Pessimistic duration estimate = 7 weeks
Most likely duration estimate = 6 weeks

Activity B: Optimistic duration estimate = 8 weeks
Pessimistic duration estimate = 12 weeks
Most likely duration estimate = 9 weeks

The probability (%) of completing the activity sequence A-B in less than 14 weeks is most nearly:

A. 2
B. 4
C. 6
D. 8

120

The results of a standard Proctor test are summarized in the figure below. A sample of soil has volume = 0.3 ft^3 and has water content = 15%.

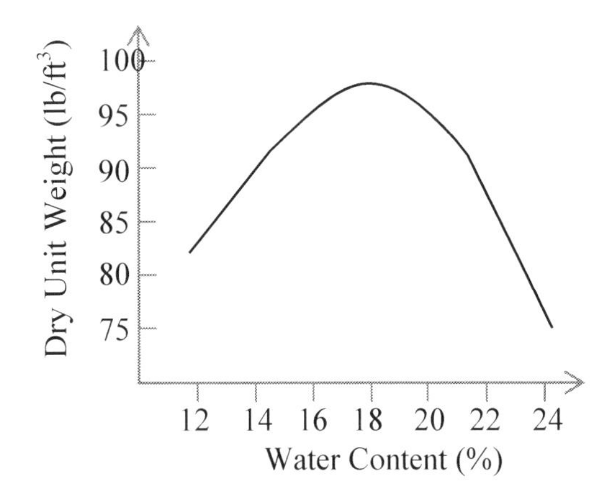

The weight of the soil sample (lb) is most nearly:

A. 32.1

B. 31.2

C. 29.6

D. 27.9

121

The table below shows cost and duration estimates for activities A-H in a project.

Activity	Normal cost ($)	Normal Duration (weeks)	Crash cost ($)	Crashed Duration (weeks)
A	3,500	7	4,500	5
B	4,500	3	5,200	2
C	12,500	4	14,500	3
D	9,000	8	11,500	6
E	6,700	9	9,300	6
F	7,200	6	8,600	5
G	9,000	7	10,400	5
H	4,500	5	5,400	4

The activity with the lowest crash cost per period ($/week) is:

A. H

B. F

C. D

D. A

122

In the layout of a site where a hazardous chemical spill has occurred and three zones – hot, warm and cold – are demarcated, which of the following statements is false?

I. The extent of the hot zone varies from site to site.
II. The hot zone should have space allocated for parking.
III. Living quarters for workers should be excluded from the hot zone only
IV. All decontamination (personnel and equipment) must occur in the hot zone only.

A. II and III only
B. III and IV only
C. II only
D. III only

123

An excavator is used to remove excess (cut) material from an embankment. The working surface for the excavator is at elevation 235.00 ft above sea level. The total volume to be excavated is 15,000 yd^3, which is distributed according the table below. The nominal productivity of the excavator is 320 yd^3/hr, which reduces by 30% when the grade difference exceeds 3 feet.

Elevation above sea level (ft)	Volume (ft^3)
230.00 – 231.00	20,800
231.00 – 232.00	41,200
232.00 – 233.00	63,400
233.00 – 234.00	34,000
234.00 – 235.00	45,600
235.00 – 236.00	67,000
236.00 – 237.00	71,200
237.00 – 238.00	38,400
238.00 – 239.00	23,400
	405,000

The total time required (hours) for the job to be completed is most nearly:

A. 42
B. 45
C. 48
D. 51

124

Dewatering at a construction site (area 34 acres) is accomplished by applying a preload at the ground surface. The resulting consolidation settlement of the underlying saturated, compressible soil is 8 inches. The expelled ground water is drained through a system of vertical drains spaced on a 10 ft by 10 ft grid. The total volume (million gallons) of groundwater collected through the drains is most nearly:

A. 3.4
B. 5.7
C. 7.4
D. 9.5

125

A roof sub-assembly 40 feet long and weighing 50,000 lb is lifted using two cranes (1 and 2) whose cables are 26 ft apart. The structure is set down onto cribbing A and B shown. The load is kept level until the load first touches down. Once the load touches down, the cable of the near crane is relieved of load and the far side crane is used to completely lower the load. The maximum load (tons) in either cable is most nearly:

A. 12.5
B. 13.5
C. 25.0
D. 27.0

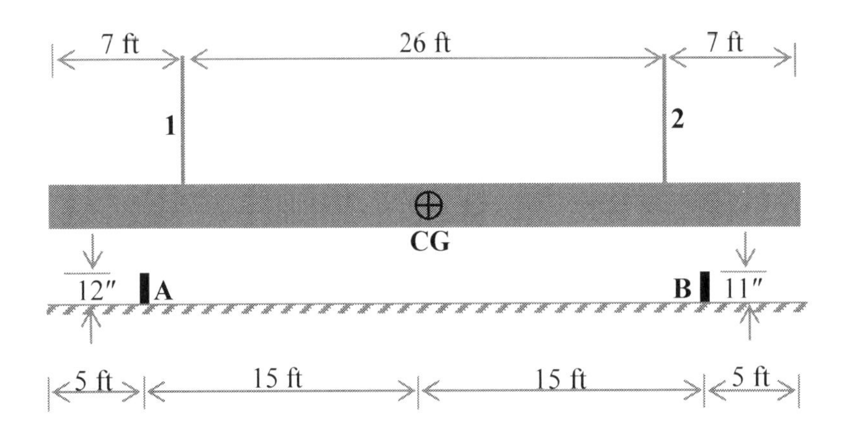

126

At a construction site, 45,000 ft^3 of soil needs to be excavated and then transported off-site. The earth is excavated by a back-hoe with a capacity of 2.4 yd^3 and a cycle time (excavation and transfer) of 5 minutes. The operational efficiency of the back-hoe is 80%. The excavated material is then transferred from the back-hoe to a fleet of small trucks that are working in tandem with the back-hoe. The capacity of each truck is 20 ft^3 and the cycle time (including load transfer, travel and dump times) is 8 min 30 secs. The efficiency of each truck is 70%.

The minimum number of trucks needed to match the production of the back-hoe is:

A. 5
B. 6
C. 7
D. 8

127

A 5/8 inch diameter steel rod (F_y = 50 ksi; F_u = 65 ksi) is used to anchor a sheet pile wall to a deadman as shown below. The force in the anchor rod is 12,000 lbs. If the minimum wall movement necessary (for active pressures to develop) is 0.7 inch, the minimum required length (feet) of the rod is most nearly:

A. 32
B. 38
C. 45
D. 51

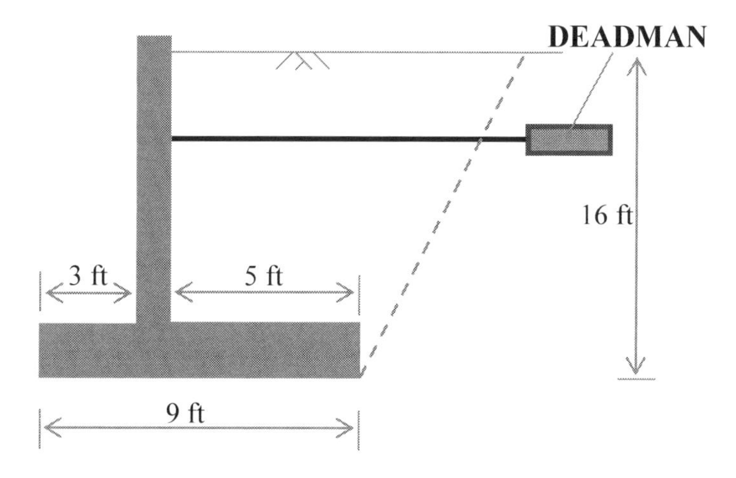

128

The project manager for an excavation project is considering costs of 4 different excavator models. The results are listed in the table below. The amount of material to be excavated is 18,000 yd^3.

Type	Production (yd^3/hr)	Fixed cost ($)	Variable cost ($/hr)
1	320	12,000	55
2	280	10,000	50
3	360	17000	55
4	300	12,000	45

The excavator which is most economical is

A. Type 1
B. Type 2
C. Type 3
D. Type 4

129

6 in x 12 in concrete cylinders were cast at the site of a reinforced concrete building. The 7-day compression test on these cylinders yielded the following failure loads (lbs): 83232, 78287, 80341, 81229 and 80452. The 7-day compressive strength (lb/in^2) of the concrete batch is most nearly:

A. 720
B. 1,440
C. 2,115
D. 2,860

130

The flow net shown below describes the seepage into a long 20 feet deep and 40 feet wide excavation made in silty sand having a coefficient of permeability (k) equal to 3 x 10^{-4} cm/sec.

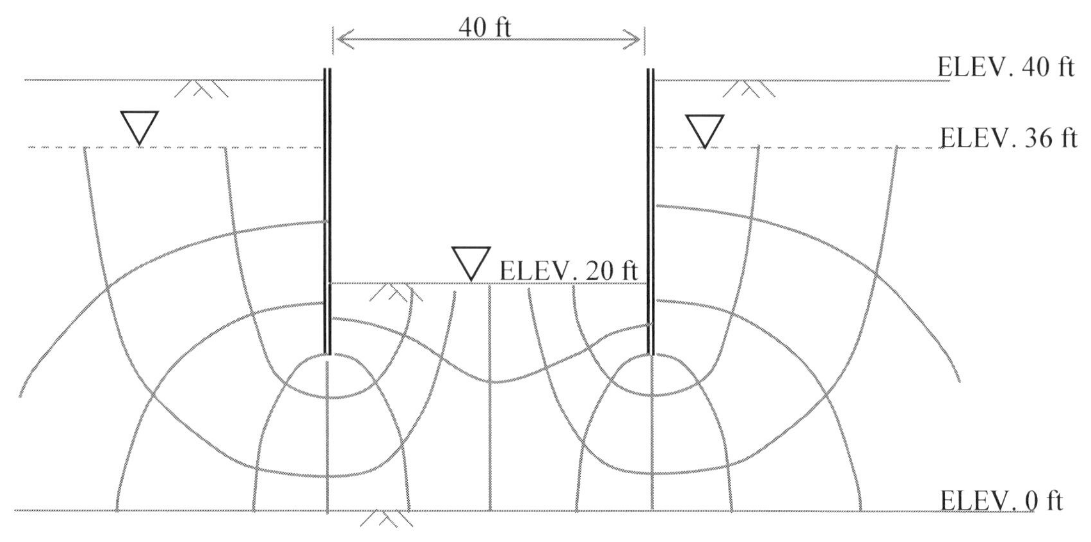

To maintain the water level at the base of the excavation the quantity of water the contractor has to pump is most nearly:

A. 3.5 x 10^{-3} ft^3/min/ft

B. 8.2 x 10^{-3} ft^3/min/ft

C. 15.0 x 10^{-3} ft^3/min/ft

D. 30.0 x 10^{-3} ft^3/min/ft

131

A free-standing wall is formed on both sides using reusable metal forms as shown. Bracing struts are spaced every 8 ft and are attached to the wall at a height of 12 ft above ground. If the maximum wind pressure acting on the wall is 40 lb/ft^2, the axial compression (lbs) in each brace is most nearly:

A. 5,410

B. 8,110

C. 12,370

D. 14,780

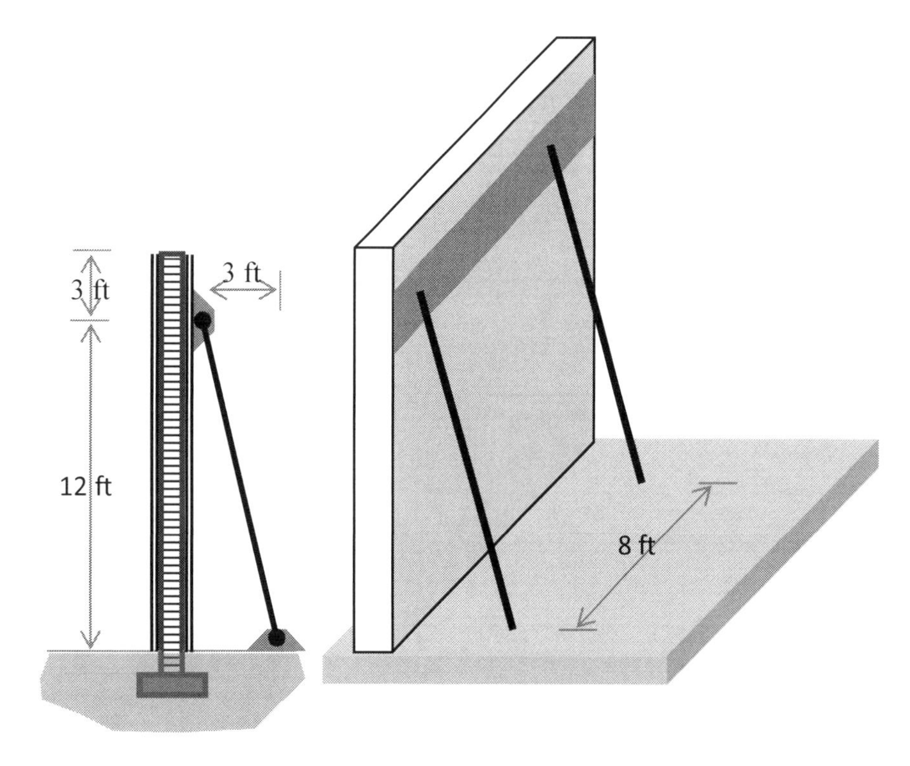

132

The table below shows cross section areas of cut and fill recorded at 5 stations spaced at 100 ft. Shrinkage factor = 0.9 and bulking factor = 1.15.

Station	Area (ft^2)	
	CUT	FILL
0 + 0.00	245.0	123.5
1 + 0.00	312.5	76.3
2 + 0.00	411.5	0.0
3 + 0.00	234.5	88.4
4 + 0.00	546.2	214.5

The net earthwork volume (yd^3) between stations 0 + 0.00 and 4 + 0.00 is most nearly

A. 3640 (cut)
B. 3780 (cut)
C. 3640 (fill)
D. 3780 (fill)

133

A masonry wall reconstruction project requires demolition of the old wall and reconstruction of new brickwork. The area of the wall is 1200 sq. ft. The demolition crew consists of 1 mason + 3 laborers having a production rate of 0.05 crew hours per sq. ft. The construction crew consists of 2 masons + 2 laborers having a production rate of 0.15 crew hours per sq. ft.

Wages are as follows:
Masons: $50/hr
Laborers: $30/hr

Total labor cost ($) for the project is most nearly:

A. 37,000
B. 41,000
C. 43,000
D. 46,000

134

A construction project has the following information in the Plan of Work:

Total duration = 113 days

Total number of tasks = 23

Number of critical tasks = 8

Total budgeted cost (labor and materials) = $1.75 million

At the end of 100 days, the following cost parameters are calculated:

Tasks completed = 15

Budgeted cost of completed tasks = $1.37 million

Expenditure on completed tasks = $1.35 million

Critical tasks completed = 5

What is the estimated actual cost to complete the project?

A. $1.698 million
B. $1.712 million
C. $1.725 million
D. $1.756 million

135

A bridge consists of a concrete slab supported by 4 equally spaced girders as shown below. The concrete deck is 36 feet wide and 9 inches thick. Steel plate girders weigh 282 lb/ft. Each girder is simply supported over a span L = 80 ft. Skew angle = 0. Repair of piers requires that a temporary jacking tower be located adjacent to each pier and a line of 6 hydraulic jacks be used to raise the girders using a jacking beam.

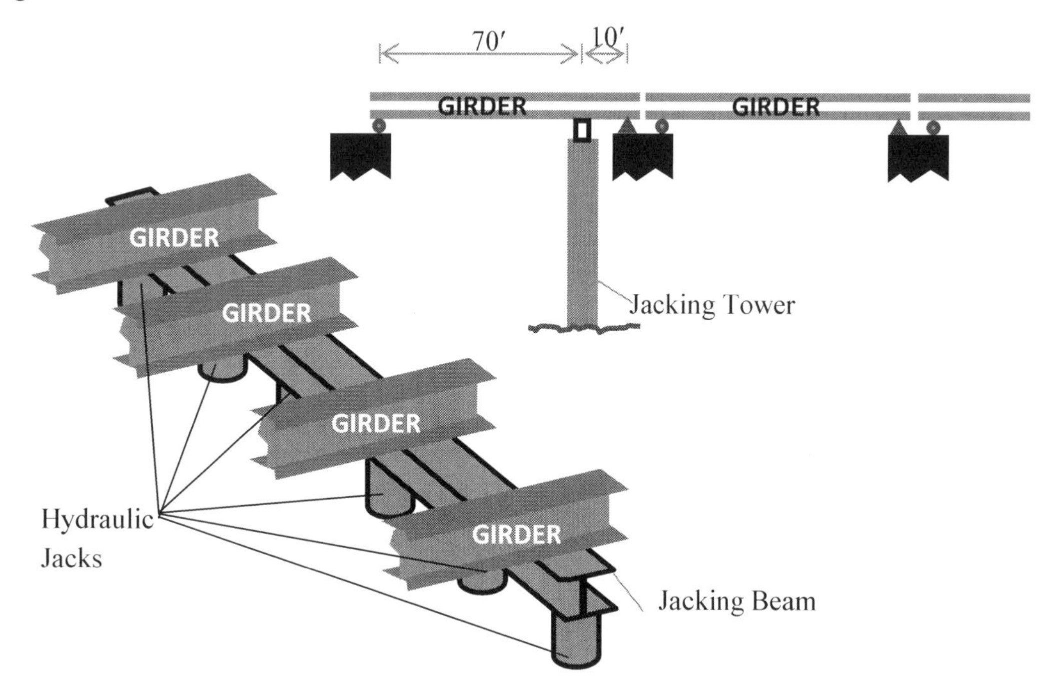

Using a factor of safety of 2.0, the minimum required capacity (tons) of the hydraulic jacks is most nearly:

A. 40
B. 32
C. 24
D. 18

136

Analysis of a suspension scaffold shows that the anticipated shear on a coupler is 1700 lb. According to OSHA criteria, the minimum design load (lbs) for the coupler is most nearly:

A. 850
B. 3,400
C. 5,100
D. 6,800

137

The figure below shows the activity on node network for the project. Numbers adjacent to the activity labels are duration in weeks. All relationships are finish to start except between B and D

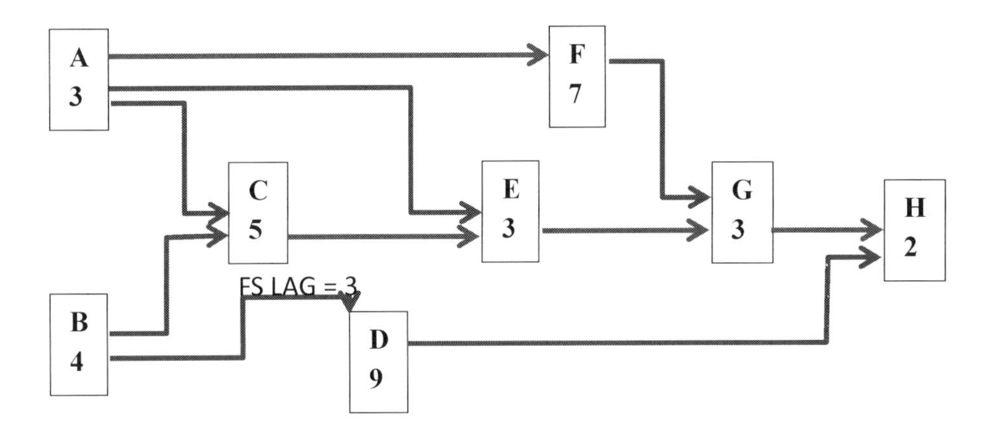

The minimum time required (weeks) to complete the project is:

A. 16
B. 18
C. 19
D. 20

138

According to the Manual of Uniform Traffic Control Devices, 2009, a temporary traffic control designated 'long-term stationary' is one that occupies a fixed work zone for a period exceeding

A. 3 days
B. 4 days
C. 5 days
D. 7 days

139

In construction management, resource leveling refers to the practice of

A. distributing costs equitably among all activities on the critical path

B. allocating resources to all activities in direct proportion to expected duration

C. eliminating resource conflicts between parallel activities

D. adjusting non-critical activities to optimize cost

140

A structure has been examined to determine a need for rehabilitation. The related costs are summarized below:

Current annual costs = $40,000

Estimated rehabilitation cost = $350,000

Annual costs projected after rehabilitation = $15,000

Expected useful life remaining = 20 years

Projected increase in residual value (at end of useful life) = $200,000

The return on investment (ROI) for performing the rehabilitation is most nearly:

A. 5%

B. 6%

C. 7%

D. 8%

THIS IS THE END OF THE CONSTRUCTION DEPTH EXAM

SOLUTIONS TO BREADTH EXAM

FOR THE

CIVIL PE EXAM

ANSWER KEY: BREADTH EXAM

001	A	009	C	017	D	025	A	033	A
002	D	010	B	018	C	026	D	034	D
003	B	011	C	019	B	027	D	035	A
004	C	012	A	020	A	028	A	036	C
005	C	013	D	021	C	029	B	037	B
006	B	014	B	022	D	030	D	038	C
007	D	015	B	023	B	031	A	039	D
008	A	016	C	024	A	032	B	040	B

Solution 001

From the table on the right, the total annual count = 225,680, from which we obtain an average monthly ADT = 18,807

Therefore, the monthly expansion factor for April = 18,807÷21,983 = 0.856

From the table on the left, the 7-day count = 123,355, of which 26,485 is weekend traffic. Therefore, the cumulative (Mon-Fri) weekday traffic is 96,870 (5 day average of 19,374.

Therefore, the daily expansion factor for Wednesday = 19,374÷19,882 = 0.974

Therefore the AAWT = 19,545x0.856x0.974 = 16,296

Answer is A

Solution 002

The easiest way to solve this problem is to use the formula for the area of a triangle:

$$A = \frac{1}{2}ab\sin C$$

where C is the angle between two sides a and b

The azimuth angles of the two lines are 65.75 and 102 degrees. Therefore the angle between them is 102 – 65.75 = 36.25

Area: $A = \frac{1}{2} \times 2345 \times 3020 \times \sin 36.25 = 2{,}093{,}798\ ft^2 = 48.067\ acres$

Answer is D

Alternatively: Arbitrarily assuming the coordinates of A to be (0, 0), the coordinates of the other two points can be found and then the method of coordinates can be used.

Solution 003

The pipe length between the two manholes has length = 1521.3 – 1062.6 = 458.70 ft.

The point of interest (low ground elevation) is located at a distance = 1305.1 – 1062.6 = 242.50 ft. from the upstream end, and 216.2 ft from the downstream end
The invert elevation at this location can be calculated by averaging the upstream and downstream invert elevations, as follows

$$238.98 \times \frac{216.2}{458.7} + 230.65 \times \frac{242.5}{458.7} = 234.58$$

Pipe wall thickness = (36.3 – 32.2)/2 = 2.05 in = 0.171 ft
Outer diameter = 36.3 in = 3.025
Elevation of the TOP of pipe = 234.58 – 0.171 + 3.025 = 237.43 ft
Soil cover = 241.55 – 237.43 = 4.12 ft

Answer is B

Solution 004

The perimeter of each gutter is 3.162 ft
Total perimeter of gutters = 6.324 ft
Surface area per mile = 6.324x5280 = 33,393.65 ft^2/mile
With a thickness of 3 inches, the volume of concrete = 8,348.4 ft^3/mile = 309.2 yd^3/mile
Cost of concrete material and placement = 232 x 309.2 = $71,734.40

Answer is C

Solution 005

Starting from A: ES_A = 0; EF_A = 0 + 5 = 5. This carries over to the successor C
For activity C: ES_C = 5; EF_C = 5 + 3 = 8
For activity B, there are two predecessors (A and C): ES_B = larger of EF_A and EF_C = 8; EF_B = 8 + 4 = 12
Since D has a single predecessor (B), ES_D = EF_B = 12. Therefore EF_D = 12 + 3 =15
E has two predecessors (B and C). Therefore ES_E = larger of EF_B and EF_C = 12. And EF_E = 12 + 5 = 17
Based on the FF lag between D and F, the EF_F = 18, based on which the ES_F = 16. However, based on EF_E = 17, the ES_F = 17. This controls.

Answer is C

Solution 006

The elevation of the point on the curve is 2.5 ft (30 in) above the crown of the sewer pipe, therefore at elev. 305.15. This point has the following offsets from the PVI:

horizontal offset h = 1230.05 – 1145.20 = +84.85 ft

vertical offset v = 305.15 – 310.56 = - 5.41 ft

The maximum length of curve (L) can be calculated from

$$\frac{L+2h}{L-2h}=\sqrt{\frac{v-G_1h}{v-G_2h}}=\sqrt{\frac{-5.41-0.05\times84.85}{-5.41--0.03\times84.85}}=\sqrt{\frac{-9.6525}{-2.8645}}=1.836$$

$$\frac{L+169.7}{L-169.7}=1.836$$

Solving L = 575.7 ft. Since this is the maximum length of curve, look for the next lower value

Answer is B

Solution 007

For ϕ = 32, active earth pressure coefficient: $K_a=\frac{1-\sin\phi}{1+\sin\phi}=0.307$

At the base of the footing (depth = 18 ft), the effective earth pressure: $K_a\gamma_{sub}H=0.307\times(126-62.4)\times18=351.5\,psf$
At the base of the footing (depth = 18 ft), the hydrostatic pressure: $\gamma_wH=62.4\times18=1123.2\,psf$
Total pressure at bottom of footing = 1474.7 psf

Total active resultant = 0.5x1474.7x18 = 13,272 lb/ft

Answer is D

Solution 008

Initial effective stress at midheight of clay layer (17 ft below surface): $p_1'=5\times115+7\times(125-62.4)+5\times(125-62.4)=1326.2\,psf$
After lowering water table, effective stress: $p_2'=12\times115+5\times(125-62.4)=1693\,psf$
Ultimate consolidation settlement: $s=\frac{HC_c}{1+e_o}\log_{10}\frac{p_2'}{p_1'}=\frac{120\times0.46}{1+0.45}\log_{10}\frac{1693}{1326.2}=4.04\,in$

Time t = 3 months = 90 days; Drainage path H_d = 5 ft (double drainage)
From the settlement time relationship, the time factor is: $t=\frac{T_vH_d^2}{c_v}\Rightarrow T_v=\frac{c_vt}{H_d^2}=\frac{0.01\times90}{5^2}=0.036$

Corresponding degree of consolidation = 22%

Therefore, after 3 months, settlement = 0.22X4.04 = 0.89 in

Answer is A

Solution 009

There are 4 loading zones on the beam – the load function on them, left to right, are w = 0, w = constant, w = 0 and w = 0 respectively. As a result, the bending moment function is M = linear, quadratic, linear and linear respectively. This eliminates choices A, B and D.

Answer is C

Solution 010

The maximum compressive stress will occur at the upper left corner of the cross section, where the uniform compression P/A will combine with the bending stress components produced by the moment about either axis (Mc/I)

$$\sigma = \frac{20 \times 10^3}{0.2 \times 0.25} + \frac{20 \times 10^3 \times 0.07 \times 0.125}{\frac{1}{12} \times 0.2 \times 0.25^3} + \frac{20 \times 10^3 \times 0.04 \times 0.1}{\frac{1}{12} \times 0.25 \times 0.2^3} = 1.552 \times 10^6\ Pa$$

Answer is B

Solution 011

With 2:1 side slopes and a depth of 5 ft, the width at the top surface = 20 + 2x10 = 40 ft
Area of flow, A = 150 ft^2
Wetted perimeter, P = 20 + 2x5x√5 = 42.36 ft
Hydraulic radius, R_h = 150/42.36 = 3.54 ft

Velocity: $V = \frac{1.486}{0.015} \times 3.54^{2/3} \times \sqrt{0.008} = 20.58\ fps$

Flow rate: Q = VA = 20.58X150 = 3087.2 cfs = 1995.2 MGD

Answer is C

Solution 012

Ultimate bearing capacity of a square footing, according to Terzaghi's theory, is given by

$$q_{ult} = 1.3cN_c + \gamma DN_q + 0.4\gamma BN_\gamma$$

From the supplied figure, for ϕ = 30°, N_c = 30, N_q = 18.5, N_γ = 22.5

$$q_{ult} = 1.3 \times 200 \times 30 + 120 \times 3 \times 18.5 + 0.4 \times 120 \times 5 \times 22.5 = 19{,}860\ psf$$

Soil pressure at base of footing = column load + soil overburden = 140,000/25 + 120X3 = 5960 psf

FS = 19,860/5,960 = 3.33

Answer is A

Solution 013

Using a tributary width of 28 inches (2.33 ft) for each stud, the uniform load acting on each stud is 30x2.33 = 70 lb/ft

Maximum bending moment in stud: $M = \frac{wL^2}{8} = \frac{70\times10^2}{8} = 875\ lb \cdot ft = 10{,}500\ lb \cdot in$

Section modulus of stud (about major axis): $S = \frac{bh^2}{6} = \frac{1.5\times5.5^2}{6} = 7.56\ in^3$

Maximum bending stress: $\sigma = \frac{M}{S} = \frac{10500}{7.56} = 1389\ psi$

Answer is D

Solution 014

In the table below, cut and fill volumes between stations are calculated using the average end area method. The net volume is calculated in the last column (positive for cut and negative for fill). The cumulative earthwork volume is the sum of the numbers in the last column.

Station	Volume (yd^3)		Net Volume (yd^3)
	CUT	FILL	
0 + 0.00			
	1032.4	370.0	+ 662.4
1 + 0.00			
	1340.7	141.3	+ 1199.4
2 + 0.00			
	1196.3	163.7	+ 1032.6
3 + 0.00			
	1445.7	560.9	+ 884.8
4 + 0.00			
			+ 3779.2

Answer is B

Solution 015

Maximum dry unit weight (Proctor) = 109 pcf
Therefore, required dry unit weight = 0.9x109 = 98.1 pcf
Weight of soil solids in embankment = 98.1x50,000 = 4.905×10^6 lb
Dry unit weight of borrow soil = 120 ÷ 1.14 = 105.3 pcf
Therefore, volume of borrow soil needed = $4.905\times10^6 \div 105.3 = 46{,}598\ ft^3 = 1{,}725.8\ yd^3$

Answer is B

Solution 016

The table below shows the precipitation in each of the three regions and the calculation of the weighted average as the average precipitation over the entire (268 acre) area

Region between	Area enclosed (acres)	Average Precipitation (inches)
1 & 2	147	0.4
2 & 3	76	0.75
3	45	0.9

The weighted average is the calculated as:

$$\bar{P} = \frac{\sum P_i A_i}{\sum A_i} = \frac{0.4 \times 147 + 0.75 \times 76 + 0.9 \times 45}{147 + 76 + 45} = 0.58\, in$$

Answer is C

Solution 017

Using a Hazen Williams roughness coefficient C = 100 and a length L= 1 mile =5280 ft

$$h_f = \frac{4.725 Q_{cfs}^{1.85} L_{ft}}{C^{1.85} D_{ft}^{4.865}} = \frac{4.725 \times 12.5^{1.85} \times 5280}{100^{1.85} \times 2^{4.865}} = 18.3\, ft$$

Answer is D

Solution 018

Volume of pond to be filled before it starts emptying = 0.78x760,000 = 592,800 gal = 79,251 ft^3
At the rate of 2 cfs, this will require 39,626 seconds = 11 hours

Answer is C

Solution 019

Taking moments about support A,

$$\sum M_A = 20 \times 10 + 10 \times 14 + 40 \times 15 + 60 \times 30 + 40 \times 45 - 60E_y = 0 \Rightarrow E_y = 75.67$$

Making a section through CD, DG and FG and then taking moments about G (see free body diagram below)

$$\sum M_G = 40 \times 15 - 75.67 \times 30 - \frac{4}{15.52} F_{CD} \times 15 - \frac{15}{15.52} F_{CD} \times 10 = 0 \Rightarrow F_{CD} = -123.43$$

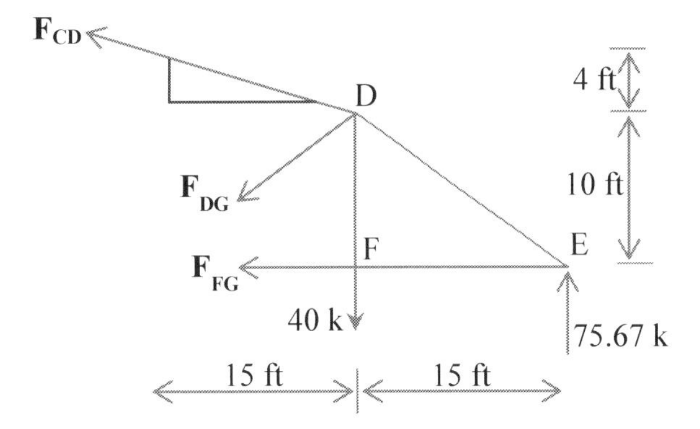

Answer is B

Solution 020

Factor of safety is given by:

$$FS = \frac{c}{\gamma H \cos^2 \beta \tan \beta} + \frac{\tan \phi}{\tan \beta}$$

For the riprap layer:

$$FS = 0 + \frac{\tan 36}{\tan 30} = 1.26$$

For the soil:

$$FS = \frac{600}{125 \times 20 \times \cos^2 30 \tan 30} + \frac{\tan 28}{\tan 30} = 0.55 + 0.92 = 1.47$$

The governing FS is 1.26

Answer is A

Solution 021

The fines fraction F_{200} = 28%. Therefore (since $F_{200} < 50$) the soil is predominantly coarse grained. First letter is S or G. Of the coarse fraction (72%), less than half (25%) is coarser than a no. 4 sieve. Therefore, the first letter is S. This eliminates A and D.

Since $F_{200} > 12\%$, second letter of the classification is determined entirely by plasticity characteristics.
PI = 45 – 21 = 24. LL = 45 and PI = 24.
This plots above the A-line. So, second letter is C.

Answer is C

Solution 022

Mass of water = 1685 – 1418 = 267 g
Volume of soil solids, V_s = 1418 ÷ 2.65 = 535 cc
Therefore, volume of voids, V_v = 855 – 535 = 320 cc
Void ratio e = V_v/V_s = 320/535 = 0.6
Answer is D

Solution 023

Buckling about the weak axis is prevented because of the bracing provided by the nails.
The Euler buckling load (about the strong axis) is given by:

$$P_e = \frac{\pi^2 EI}{L^2} = \frac{\pi^2 \times 1.5 \times 10^6 \times \frac{1}{12} \times 1.5 \times 5.5^3}{(12 \times 12)^2} = 14848\ lb = 14.85\ kips$$

Answer is B

Solution 024

Using the reservoir surface as point 1 and open end of pipe as point 2, both of these points are at atmospheric pressure $p_1 = p_2 = p_{atm}$. Also, since the reservoir is 'large', by the continuity principle, $V_1 \approx 0$

Head loss (friction) in the pipe, using the Darcy-Weisbach equation, is

$$h_f = f\frac{L}{D}\frac{V^2}{2g} = 0.024 \times \frac{500}{2} \times \frac{V^2}{2 \times 32.2} = 0.0932V^2$$

Writing Bernoulli's equation between points 1 and 2

$$\frac{p_{atm}}{\gamma} + 324.5 + 0 - 0.0932V^2 = \frac{p_{atm}}{\gamma} + 295.8 + \frac{V^2}{2 \times 32.2} \Rightarrow 0.109V^2 = 28.7 \Rightarrow V = 16.2\ fps$$

Flow rate: Q = VA = 50.98 cfs

Answer is A

Solution 025

Time of overland flow = $t_s + t_d$ = 5 + 13 = 18 minutes
Channel travel time = 1200 ÷ 5 = 240 seconds = 4 minutes
Time of concentration for point B = 18 + 4 = 22 min
For duration = 22 min and return period = 20 years, intensity = 1.5 in/hr

Answer is A

Solution 026

The tangent length (calculated from PI, PC pair – can also be calculated from PI-PT pair)

$$T = \sqrt{(1250 - 100)^2 + (320 - 817)^2} = 1252.8$$

The azimuth angle of the back tangent (calculated as inverse tan of departure divided by latitude)

$$Az_{BT} = \tan^{-1}\left(\frac{817 - 320}{100 - 1250}\right) = 156.63$$

The azimuth angle of the forward tangent (calculated as inverse tan of departure divided by latitude)

$$Az_{FT} = \tan^{-1}\left(\frac{900 - 817}{1350 - 100}\right) = 3.80$$

Therefore, the deflection angle I = 3.80 – 156.63 = – 152.83 (negative sign means deflecting left)

Since T = R tan (I/2), solving for R = 302.724 ft

Degree of curve D = 5729.578/R = 18.93 degrees

Answer is D

Solution 027

Height of warehouse = 26 ft
Perimeter = 2x(40+110) = 300 ft
Surface area of 4 walls = 300x26 = 7800 sq. ft.
Inclined length of roof = $2\sqrt{(8^2 + 20^2)}$ = 43.08 ft
Surface area of roof = 110x43.08 = 4739 sq. ft.
Area of 4 openings to be subtracted = 4x8x13 = 416 sq. ft.
Total area of sheathing = 7800 + 4739 – 416 = 12,123 sq. ft.

Answer is D

Solution 028

Collapsing the 3D structures into a 2D one, and representing the *total* pad reaction on the left (2 pads) as R_L and the *total* pad reaction on the right (2 pads) as R_R, taking moments about R_L, we get:

$$9000 \times 14 + 800 \times 37 - R_R \times 36 + 8000 \times 56 = 0 \Rightarrow R_R = 16,767\ lb$$

With the load on the right, the maximum compression reaction on the ground will occur on the pads on the right.
Reaction on each pad = 8,384 lb

Soil pressure under the pads on the right (area = 7.07 ft^2) = 1,186 lb/ft^2

Answer is A

Solution 029

For concrete beams, the bending moment due to loading determines the extent of cracking experienced by the beam. The effective moment of inertia, used for calculating deflections according to the elastic theory, is between I_{cr}, the cracked moment of inertia and I_g, the gross moment of inertia. Only statement IV is correct.

Answer is B.

Solution 030

By using the 0.2% offset method (drawing a line parallel to the initial tangent through the strain offset = 0.2% = 2000 $\mu\varepsilon$), the yield stress is 38 ksi

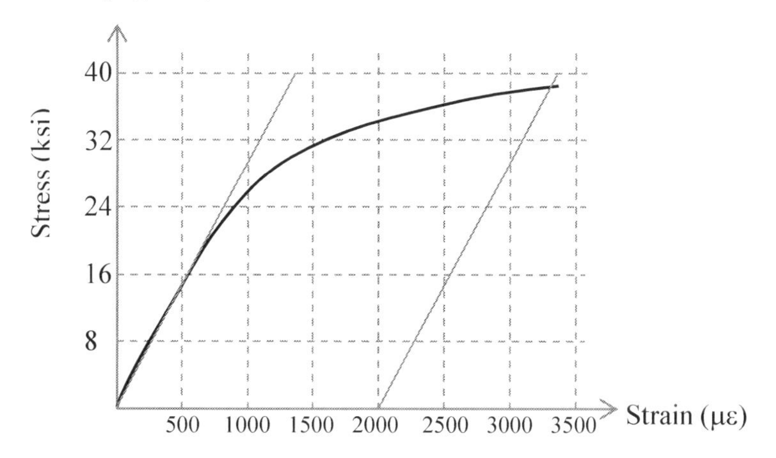

Answer is D

Solution 031

The thickness of sand that is affected by the pressure from the mat foundation is 15 – 3 = 12 ft = 180 in
Assuming that the mat exerts a uniform pressure of 600 psf on the entire layer,
Vertical strain = Vertical stress/E = $600/1\text{x}10^6 = 6\text{x}10^{-4}$
Vertical displacement = $180\text{x}6\text{x}10^{-4}$ = 0.108 in

Answer is A

Solution 032

II is incorrect. LS – EF – D = LS – (ES + D) – D = LS – ES – 2D = TF – 2D, which can equal TF only if duration D = 0 (which is a trivial solution)
Answer is B

Solution 033

In the long term, clay soils go through consolidation due to the expulsion of pore water. Results obtained from the CD (consolidated drained) triaxial test are good predictors of long term conditions. Therefore, statement I is correct.
The CD test involved allowing pore water to slowly drain from the soil and is therefore slower than the UU test. Statement II is correct and statement III is false. During the CD test, pore pressures are not allowed to build up. Statement IV is incorrect.
Answer is A

Solution 034
Material delivery = 540 yd^3/hr (loose soil), which is equivalent to 540x0.83 = 448.2 yd^3/hr compacted
Roller covers ground at 3 mph x 8 ft = 126720 ft^2/hr. 0.5 ft thick layer gets compacted in 4 passes. Therefore each pass compact the equivalent of 0.125 ft, which means it compacts 15,840 ft^3 (587 yd^3) of soil per pass. This is ideal capacity. Working 50 minutes per hour, roller compacts 50/60x587 = 489 yd^3/hr. Therefore, only 1 roller is needed to handle the delivery of the material.

Answer is D

Solution 035
Midspan deflection of a simply supported beam with point load is given by:

$$\Delta_{max} = \frac{PL^3}{48EI} = \frac{20 \times (24 \times 12)^3}{48 \times 29000 \times 1490} = 0.23\ in$$

Answer is A

Solution 036

The total vertical force (weight of concrete 3 & 4) + weight of soil (1) = 20,600 lb/ft
Horizontal friction force that can be mobilized under the footing = 0.6x20,600 = 12,360 lb/ft

$$FS = \frac{12{,}360}{5{,}400} = 2.29$$

Answer is C

Solution 037

The total volume of runoff that collects in the detention pond = 120 acre x 1 inch = 120 ac-in = 4.356x10^5 ft^3 = 12,333 m^3 = 1.2x10^7 L
Total mass of sediment = 6.167x10^7 g = 6.167x10^4 kg
Bulk specific gravity of the sediment = 80/62.4 = 1.282
Density of sediment = 1282 kg/m^3
Volume occupied by sediment = 6.167x10^4 kg ÷ 1282 kg/m^3 = 48.1 m^3 = 1699 ft^3
Depth occupied by sediment = 1699 ÷ (2x43560) = 0.0195 ft = 0.234 inch
Answer is B

Solution 038

Breaking strain calculated as: Elongation at break ÷ Sample initial length
The values calculated for samples 1-5 are: 0.0728, 0.0703, 0.0719, 0.0716 and 0.0709
The average breaking strain = 0.0715 = 71,500 $\mu\varepsilon$

Answer is C

Solution 039

Depth ratio d/D = 30/48 = 0.625. For this depth ratio, for constant n, $V/V_f = 1.0857$

Longitudinal slope: $S = \frac{365.82-354.28}{1200} = 0.0096$
Velocity for pipe flowing full: $V_f = \frac{0.590}{n} D^{2/3} S^{1/2} = \frac{0.590}{0.014} \times 4^{2/3} \times 0.0096^{1/2} = 10.4\ cfs$

Actual velocity in pipe (when d = 30 in) is V = 11.3 fps

Answer is D

Solution 040
Underpinning (I) is commonly used to support existing structures to counter the possible loss of bearing support from adjacent excavation. Slurry walls (III) can be used to 'isolate' sensitive structures from construction activities.

Answer is B

SOLUTIONS TO CONSTRUCTION EXAM

FOR THE

CIVIL PE EXAM

ANSWER KEY: CONSTRUCTION DEPTH EXAM

101	B
102	A
103	C
104	D
105	A
106	D
107	C
108	C

109	B
110	D
111	A
112	C
113	D
114	A
115	B
116	C

117	A
118	B
119	D
120	A
121	D
122	B
123	D
124	C

125	B
126	C
127	C
128	B
129	D
130	B
131	C
132	A

133	A
134	C
135	A
136	D
137	B
138	A
139	D
140	B

Solution 101

Stations 0+00 and 12+00 are match points on the mass diagram
Hauling zone no. 1: cut from stations 3+30 and 5+00 to the fill from 0+00 to 3+30
Hauling zone no. 2: cut from stations 5+00 and 6+10 to the fill from 11+00 to 12+00
Hauling zone no. 3: cut from stations 9+00 and 11+00 to the fill from 6+10 to 9+00

Answer is B

Solution 102

There are 32 nodes in all, of which 6 are associated with a ¼ cell, 12 with a ½ cell, 2 with a ¾ cell and 12 with a full cell. Each cell has area = 100x100 ft^2. Assigning the appropriate weights (area) to these values, the volume can be calculated as:

$$V = 100 \times 100 \times \left[\frac{1}{4} \times (2.14 + 1.65 + 3.27 + 2.25 + 2.65 + 2.64) + \frac{1}{2} \times (3.45 + 2.80 + 2.88 + 3.12 + 3.12 + 2.70 + 3.20 + 4.65 + 3.44 + 2.15 + 2.64 + 3.00) + \frac{3}{4} \times (3.12 + 2.77) + 1 \times (2.32 + 1.65 + 2.77 + 3.25 + 3.44 + 2.70 + 3.25 + 3.44 + 2.70 + 3.00 + 3.85 + 3.15)\right] = 621{,}625\ ft^3 = 23{,}023\ yd^3$$

Answer is A

Solution 103

The swell potential = 15% means that 1 ft^3 bank soil is equivalent to 1.15 ft^3 loose soil. Therefore 2.4 yd^3 loose soil is equivalent to 2.087 yd^3 bank soil.

Production = 2.087 yd^3 (bank) per 4.5 minutes = 27.83 yd^3/hr
Answer is C

Solution 104

N = blows per inch = 7/4 = 1.75

$$R_u = 6.6 \times 0.7 \times 0.8 \times 8 \times \ln(10 \times 1.75) = 120.9\ kips$$

Answer is D

Solution 105

A and B are initial activities (with no predecessors). Therefore, ES_A and ES_B = 0

EF_A = 0 + 3 = 3

EF_B = 0 + 4 = 4

Since C has two predecessors (A & B), ES_C = larger of EF_A and EF_B = 4. Therefore, EF_C = 4 + 5 = 9

D has a single predecessor (B). Therefore, ES_D = EF_B = 4. EF_D = 4 + 9 = 13

Since E has two predecessors (A & C), ES_E = larger of EF_A and EF_C = 9. Therefore, EF_E = 9 + 3 = 12

F has a single predecessor (A). Therefore, ES_F = EF_A = 3. EF_F = 3 + 7 = 10

Since G has two predecessors (E & F), ES_G = larger of EF_E and EF_F = 12.

Answer is A

Solution 106

The weld appears on both sides (arrow side and far side). Therefore I is incorrect.

The fillet weld symbol (triangle) in figure III is flipped (vertical edge on the right). Therefore, it is incorrect.

The weld does not appear to be an all-around weld. Therefore IV is incorrect.

Answer is D

Solution 107

Assume weight of cement = 100 lb	Volume of fine aggregate = 100/(3.15x62.4) = 0.5088 ft^3
Weight of fine aggregate = 170 lb	Volume of fine aggregate = 170/(2.6x62.4) = 1.0478 ft^3
Weight of coarse aggregate = 250 lb	Volume of fine aggregate = 250/(2.7x62.4) = 1.4839 ft^3
Weight of water = 40 lb	Volume of water = 40/62.4 = 0.6410 ft^3

Volume (excluding air) = 3.6815 ft^3, which represents 97% of total volume. Therefore, total volume = 3.7954 ft^3

Unit weight = total weight/total volume = 560/3.7954 = 147.55 lb/ft^3

Specific gravity = 147.55/62.4 = 2.365

Answer is C

Solution 108

The two methods recognized by ASTM C 1074 are the Nurse Saul method and the Arrhenius function method.
Answer is C

Solution 109

For column forms, with a rate of pour R = 8 ft/hr and T = 70°F, the base value of the lateral pressure, p = 1090 lb/ft^2
Type II cement with no retarders, $C_c = 1.0$
For unit weight = 130 pcf, $C_w = 0.5\left(1+\frac{\gamma}{145}\right) = 0.5\left(1+\frac{130}{145}\right) = 0.95$

Maximum lateral pressure = 1.0x0.95x1090 = 1036 psf, but not to exceed γh. At midheight (concrete depth = 7 ft), this maximum pressure = 130x7 = 910 psf. This value controls.

Answer is B

Solution 110
The total load of D + 0.3D + 0.1D from slab no. 5 is distributed equally through interconnected slabs 2, 3 and 4. Thus, each of these 3 floors carries 1.4D/3 = 0.47D.
The total load transmitted to the reshores below level 4 = D + 0.3D + 0.1D + D + 0.05D = 2.45D, of which slab 4 carries 1.47D. Therefore, the load transmitted down to the reshores below level 4 = 2.45D – 1.47D = 0.98D

Answer is D

Solution 111

Uniform load tributary to each soldier pile = 470x8 = 3760 lb/ft
Answer is A

Solution 112

According to Table D-3 in 1926.56(a), the minimum illumination intensity for carpentry shops is 10 foot-candles.

Answer is C

Solution 113

Statement I is true. Stations 6+20 and 11+00 have the same ordinate (1000 yd^3), implying that the net earthwork between these stations is zero

Statement II is not true. Station 5+00 has a zero ordinate, implying that the net earthwork between station 0+00 and 5+00 is zero
Statement III is true. Between stations 6+20 and 9+00, the mass diagram goes down, implying a fill, while between stations 9+00 and 11+00, the mass diagram goes up, implying a cut

Answer is D

Solution 114

Statement B is true. Seismic reflection methods are useful for identifying deep features, but are of limited use in shallow ground investigation.
Statement C is true. Gravity surveys are conducted by measuring minute variations in the earth's gravitational force. This requires very delicate and expensive equipment.
Statement D is true. Resistivity surveys are of limited use in shallow ground investigation because of the difficulty in interpreting the results.

Answer is A

Solution 115

1926 Subpart P, Appendix A: Soil with unconfined compression strength greater than 1.5 tons/ft^2 (3000 lb/ft^2) is classified as type A unless it is subject to vibration (and some other restrictive conditions) in which case it is classified as type B

Answer is B

Solution 116

The cycle time for the excavator = excavation time + 2 way travel time over 500 ft at 10 mph = 2 min + 1000 ÷ (10x1.47) seconds = 68 sec + 2 min = 3.13 min
Production rate for excavator = 2.44 yd^3 ÷ 3.13 min = 46.8 yd^3/hr

The cycle time for each truck = 2 way travel time over 3800 ft at 15 mph + loading time = 7600 ÷ (15x1.47) seconds + 30 seconds = 344.7 sec + 0.5 min = 6.24 min
Production rate for each truck = 50 ft^3 ÷ 6.24 min = 8 ft^3/min = 0.30 yd^3/min = 17.8 yd^3/hr

Number of trucks needed = 46.8 ÷ 17.8 = 2.63. At least 3 trucks are needed.

Answer is C

Solution 117

Total length of reinforcement in edge strips (No. 5 bars top and bottom) = 12x14 + 12x18 = 384 ft
Total length of midspan reinforcement in long direction (No. 4 bars top and bottom) = 12x9 = 108 ft

Total length of midspan reinforcement in short direction (No. 6 bars top and bottom) = 12x7 = 84 ft

Total weight of steel = 384x1.04 + 108x0.67 + 84x1.50 = 597.72 lb

Answer is A

Solution 118

Taking the lower right corner of the building (ground level) as the origin (0, 0), the foot of the crane boom is at B (40, 10), upper right corner of the building (roof level) is at (0, 50) and point A is at (6, 56). Length of crane boom between boom foot and point A is calculated using Pythagoras' formula:

$$BA = \sqrt{(6-40)^2 + (56-10)^2} = 57.2\,ft$$

Since the boom length is 110 ft, using similar triangles, the horizontal projection of BC is related to the horizontal projection of BA (34 ft) according to:

$$x_{BC} = \frac{110}{57.2} \times 34 = 65.38\,ft$$

Desired distance X = 65.38 – 40 = 25.38 ft

Answer is B

Solution 119

Mean duration for activity A: $\mu_A = \frac{a+4m+b}{6} = \frac{4+4\times6+7}{6} = 5.83$

Standard deviation for activity A: $\sigma_A = \frac{b-a}{6} = \frac{7-4}{6} = 0.5$

Mean duration for activity B: $\mu_B = \frac{a+4m+b}{6} = \frac{8+4\times9+12}{6} = 9.33$

Standard deviation for activity B: $\sigma_B = \frac{b-a}{6} = \frac{12-8}{6} = 0.67$

Expected mean duration for A-B: $\mu_{AB} = 5.83 + 9.33 = 15.17$

Variance for A-B: $\sigma_{AB}^2 = \sigma_A^2 + \sigma_B^2 = 0.5^2 + 0.67^2 = 0.69 \Rightarrow \sigma_{AB} = 0.83$

Therefore, sequence AB can be modeled as normal distribution with mean, μ = 15.17 and standard deviation σ = 0.83

Z score for X = 14 is $Z = \frac{X-\mu}{\sigma} = \frac{14-15.17}{0.83} = -1.4$

Cumulative area (from the normal distribution tables) to Z = -1.4 is 0.0808. Thus, the probability that Z < -1.4, which implies X < 14, is 0.0808, or 8.1%

Answer is D

Solution 120

At w = 0.15, the dry unit weight, γ_d = 93 lb/ft^3

Total unit weight $\gamma = \gamma_d(1+w)$ = 106.95 lb/ft^3
Soil sample weight = 0.3x106.95 = 32.09 lb

Answer is A

Solution 121

The cost to crash per period is given by: $\frac{Crash\ cost - Normal\ cost}{Normal\ time - crash\ time}$

Activity	$\frac{CC - NC}{NT - CT}$
A	1000/2 = 500
B	700/1 = 700
C	2000/1 = 2,000
D	2,500/2 = 1,250
E	2,600/3 = 867
F	1,400/1 = 1,400
G	1,400/2 = 700
H	900/1 = 900

A is the activity with the lowest crash cost per period.

Answer is D

Solution 122

Statement I is true. The radius of the hot zone (closest to the spill location) depends on the toxicity of the chemical. It can also have directional variation, depending on local conditions such as prevailing wind direction etc.

Statement II is true because even though majority of vehicles are parked in the warm zone, there should be some provision for parking vehicles and equipment in the hot zone so that they can be decontaminated before leaving the hot zone.

Statement III is not a true statement. Living quarters should be confined to the cold zone, preferably upwind from the hot zone.

Statement IV is not a true statement. Long term decontamination activities should take place in the warm zone.

Answer is B

Solution 123

All volumes that exist between elevation 232.00 – 238.00 (235 ± 3) are removed at the nominal rate of 320 yd^3/hr, while those existing outside this interval are removed at 224 yd^3/hr

Volume outside the elevation interval 235 ± 3 = 85,400 ft^3 = 3,163 yd^3

Time required = 3163 ÷ 224 = 14.1 hrs

Volume within the elevation interval 235 ± 3 = 405,000 – 85,400 = 319,600 ft^3 = 11,837 yd^3

Time required = 11837 ÷ 320 = 37.0 hrs
Total time required to complete job = 51.1 hours

Answer is D

Solution 124

The consolidation settlement is due to the expulsion of water from the voids. The volume change of the compressible soil layer is therefore, equal to the volume of water expelled.
Volume = 34 acres x 8 inches = 272 ac-in = 22.67 ac-ft = 987,360 ft^3 = 7.39 million gallons

Answer is C

Solution 125

Evidently, until the load touches down at either end, the load will symmetrically distribute equally to the cables 1 and 2. At that stage, each cable will carry half of 50,000 lb = 25,000 lb.
Because of the unequal height of the cribbing at A and B, the load will touch down at A first. According to the given protocol, the near side crane 1 will be deactivated (cable tension will become zero) and the far cable (crane 2) will be used to lower the load onto B. During this stage, the distance from the CG to cribbing A is 15 ft and the distance from the CG to cable 2 is 13 ft.

Load in cable 2 = 15/28x50,000 = 26,786 lb = 13.4 tons
Answer is B

Solution 126

Production of the back hoe: $P_{BH} = \frac{0.8\times2.4}{5} = 0.384\ \frac{yd^3}{min} = 10.368\frac{ft^3}{min}$
Production of a single truck: $P_T = \frac{0.7\times20}{8.5} = 1.647\ \frac{ft^3}{min}$
Number of trucks needed: $n = \frac{P_{BH}}{P_T} = \frac{10.368}{1.647} = 6.3$
Therefore, at least 7 trucks are needed to match the production of the back-hoe.

Answer is C

Solution 127

Average normal stress in rod: $\sigma = \frac{P}{A} = \frac{12000}{\frac{\pi}{4}\times\left(\frac{5}{8}\right)^2} = 39{,}114\ psi$
Since this is less than the yield stress, we can assume that elastic Hooke's law applies.

Elongation: $\Delta = \frac{PL}{AE} = \frac{\sigma L}{E} \Rightarrow L = \frac{E\Delta}{\sigma} = \frac{29\times10^6\times0.7}{39{,}114} = 519\ in = 43.3\ ft$

Answer is C

Solution 128

The time needed by each excavator can be calculated by diving 18,000 yd^3 by the production rate (yd^3/hr). This will yield the 'time to complete' in hours, which can then be multiplied by the variable cost rate ($/hr) to obtain the variable cost. The fixed cost can then be added in.

Total cost = Fixed Cost + 18,000 yd^3 ÷ Production (yd^3/hr) x Variable cost ($/hr)

The total cost for types 1-4 are $15,094, $13,214, $19,750 and $14,700 respectively.

Answer is B

Solution 129

Cross-sectional area: $A = \pi r^2 = \pi \times 3^2 = 28.27\ in^2$

Average compressive load, P = 80,708 lb
Compressive strength = 80,708 ÷ 28.27 = 2,855 psi

Answer is D

Solution 130

$K = 3 \times 10^{-4}$ cm/sec = 1×10^{-5} ft/sec
For each side, $N_f = 3$, $N_e = 7$, H = 16 ft
Flow rate (per unit length) $q = 1 \times 10^{-5} \times 3/7 \times 16 = 6.86 \times 10^{-5}\ ft^3/s = 4.11 \times 10^{-3}\ ft^3/min$
Therefore, total seepage into the trench = $2 \times 4.11 \times 10^{-3}\ ft^3/min/ft = 8.2 \times 10^{-3}\ ft^3/min/ft$

Answer is B

Solution 131

The tributary width for each strut being 8 ft, the wind load of 40 psf can be converted to a line load of 320 lb/ft. Assuming a hinged base, we have the free body diagram shown on the right

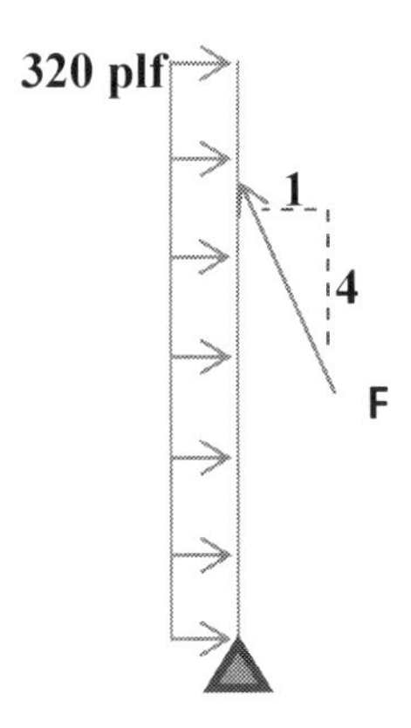

$$M_{base} = 320 \times 15 \times 7.5 - \frac{1}{\sqrt{17}} F \times 12 = 0$$

F = 12,369 lb

Answer is C

Solution 132

In the table below, cut and fill volumes between stations are calculated using the average end area method. The fill quantities are then adjusted by dividing by the shrinkage factor of 0.9. The net volume is calculated in the last column (positive for cut and negative for fill). The cumulative earthwork volume is the sum of the numbers in the last column.

Station	Volume (yd^3)			Net Volume (yd^3)
	CUT	FILL	ADJUSTED FILL	
0 + 0.00				
	1032.4	370.0	411.1	+ 621.3
1 + 0.00				
	1340.7	141.3	157.0	+ 1183.7
2 + 0.00				
	1196.3	163.7	181.9	+ 1014.4
3 + 0.00				
	1445.7	560.9	523.3	+ 822.5
4 + 0.00				
				+ 3641.9

Answer is A

Solution 133

Demolition crew: Labor cost per crew hour = 1x50+3x30 = \$140/crew hour

Construction crew: Labor cost per crew hour = 2x50+2x30 = \$160/crew hour

Demolition cost = 1200 ft^2 x 0.05 C.H/ft^2 x \$140/C.H. = \$8,400

Construction cost = 1200 ft^2 x 0.15 C.H/ft^2 x \$160/C.H. = \$28,800

Total cost = \$37,200

Answer is A

Solution 134

Actual cost of work performed, ACWP = \$1.35 million

Budgeted cost of work performed, BCWP = \$1.37 million

Cost performance index, CPI = 1.37/1.35 = 1.0148

BAC = original project estimate = \$1.75 million

Estimate to complete (millions of dollars), ACP = ACWP + (BAC – BCWP)/CPI = \$1.35 + (1.75 – 1.37)/1.0148 = 1.724

Answer is C

Solution 135

The total dead load of the bridge deck + girders = 36x80x0.75x150 + 4x80x282 = 324,000 + 90,240 = 414,240 lb. This force acts at midspan.
Taking moments about the support far from the jacked end, total force on jacks = 414 k x 40/70 = 237 k
Assuming this load is equally shared by the 6 jacks, each jack needs to exert a force of 39.5 k = 19.7 tons

If a minimum FS = 2.0 is desired, the jacks must be rated at 39.5 tons

Answer is A

Solution 136

Each scaffold and scaffold component shall be capable of supporting, without failure, its own weight and at least four times the maximum intended load applied or transmitted to it.

1700 x 4 = 6800 lb

Answer is D

Solution 137

A and B are initial activities (with no predecessors). Therefore, ES_A and ES_B = 0
$EF_A = 0 + 3 = 3$
$EF_B = 0 + 4 = 4$
Since C has two predecessors (A & B), ES_C = larger of EF_A and EF_B = 4. Therefore, $EF_C = 4 + 5 = 9$
D has a single predecessor (B). However, the finish-to start lag means that D cannot start until 4 + 3 = 7.
Therefore, $ES_D = EF_B + 3 = 7$. $EF_D = 7 + 9 = 16$
Since E has two predecessors (A & C), ES_E = larger of EF_A and EF_C = 9. Therefore, $EF_E = 9 + 3 = 12$
F has a single predecessor (A). Therefore, $ES_F = EF_A = 3$. $EF_F = 3 + 7 = 10$

Since G has two predecessors (E & F), ES_G = larger of EF_E and EF_F = 12. EF_G = 12 + 3 = 15

Since H has two predecessors (D & G), ES_H = larger of EF_D and EF_G = 16. EF_H = 16 + 2 = 18

Answer is B

Solution 138

Answer is A (MUTCD section 6G.02)

Solution 139

A is incorrect. The critical path is what it is. Cost adjustment on the critical path only does not address resource demand elsewhere.

B is incorrect. Usually, allocating more resources to an activity will reduce the duration. Also, this option says nothing about the 'simultaneity' of resource heavy activities.

C could be correct, except that a complete elimination of conflict between parallel activities is not always feasible.

Of these choices, D is the closest to the concept of resource leveling. Since we have no choice of moving the critical activities, the non-critical activities have to be adjusted (time occupied on project timeline), so that the demand on resources is as close to uniform as possible, thereby reducing losses due to idle time.

Answer is D

Solution 140

At the rate of return (i), the present worth should be zero.

The \$350k capital expenditure is a present value (P). NEGATIVE
The \$25k reduction in annual costs is an annuity (A) POSITIVE
The \$200k increase in salvage value is a future sum (F) POSITIVE

Converting all of these to present worth, the net present worth can be written:

$$PW = -350 + 25\left(\frac{P}{A}, i, 20\ yrs\right) + 200\left(\frac{P}{F}, i, 20\ yrs\right) = 0$$

For i = 5%, PW = 37k
For i = 6%, PW = -0.9k Actual answer 5.97%
For i = 7%, PW = -33.5k

Answer is B

Made in the USA
San Bernardino, CA
22 June 2016